Fun Physics Projects for Tomorrow's Rocket Scientists

About the Author

Alan Gleue is a physics teacher and a science department chairperson at Lawrence High School (LHS) in Lawrence, Kansas, and has many opportunities to engage students in the creative process of science and engineering. In his physics courses, the students study motion, thermodynamics, forces, electricity, and optics with many opportunities to complete projects and activities. He started teaching at LHS in 1996; prior to that, he taught high school in Ohio and junior high in Kansas City, Kansas. He has won several teaching awards and often participates in extracurricular workshops to improve his teaching and gain ideas to bring into his classroom. Although Alan teaches physics exclusively now, he has experience teaching chemistry, astronomy, and physical science. He studied at Kansas State University, graduating with a degree in engineering and then completed his master's degree at the University of Kansas in curriculum and instruction.

Fun Physics Projects for Tomorrow's Rocket Scientists

A Thames & Kosmos Book

Alan Gleue

McGraw
Hill

New York Chicago San Francisco
Lisbon London Madrid Mexico City
Milan New Delhi San Juan
Seoul Singapore Sydney Toronto

The McGraw·Hill Companies

Cataloging-in-Publication Data is on file with the Library of Congress

Fun Physics Projects for Tomorrow's Rocket Scientists: A Thames & Kosmos Book

1234567890   QDB QDB   1098765432

ISBN   978-0-07-179899-0
MHID   0-07-179899-4

Sponsoring Editor: Roger Stewart

Editorial Supervisor: Jody McKenzie

Project Editor: LeeAnn Pickrell

Acquisitions Coordinator: Molly Wyand

Copy Editor: LeeAnn Pickrell

Proofreader: Paul Tyler

Indexer: Rebecca Plunkett

Production Supervisor: Jean Bodeaux

Composition: Cenveo Publisher Services

Illustration: Greg Scott, Cenveo Publisher Services, and Jay's Publisher's Services

Art Director, Cover: Jeff Weeks

Cover Designer: Jeff Weeks

Series Design: Mary McKeon

To my three boys: do you remember our *tennis ball cannon?*

Contents

Introduction

If you happen to open a physics textbook, you will see chapters about motion, forces, sound and light, heat, lenses and mirrors, and electricity. You might find a textbook with lots of strange symbols and complicated-looking math equations. Not to worry! In this book, we focus mainly on the physics of motion. The title is, after all, *Fun Physics Projects for Tomorrow's Rocket Scientists*. By studying motion, you'll learn much that the rocket scientist needs to know.

Everything in the universe is in motion. As you read this book, you may be sitting in your room and feel that you are perfectly still. But the Earth, which you are on, is spinning on its axis at the rate of just over 1,000 miles per hour (at the equator), while it also revolves around the sun at a rate of 67,000 miles per hour. The sun itself is moving through the Milky Way galaxy. When you talk about an object being still, or at rest, it is only still in relation to the world around it.

In this book, I'm going to walk you through performing experiments that help you understand many aspects of the physics of motion, including constant speed, acceleration, free fall, and more. To do this, you'll use the techniques of science:

- Making observations
- Composing hypotheses
- Devising experiments
- Taking measurements
- Performing calculations
- Formulating conclusions

You can't really *do* physics without doing a little math, but I'll keep it simple and straightforward.

NOTE

The website for this book contains videos from my experiments and other handy tools. Go to http://www.mhprofessional.com/fun_physics to access the site.

You can measure motion with a stopwatch and a meter stick. In this book, however, I recommend using a couple of tools that will not only make the experiments more fun but also enable you to be more accurate.

The first tool is a video camera. Any kind of digital video camera will do. If you have access to a Flip Video camera, that would be perfect. You may have a digital still camera that takes videos. Or you may have a smartphone, such as an iPhone, that takes videos. I also recommend a tripod or some sort of still mount to put the camera on.

The second tool is actually a piece of software. Throughout the book, I use the Tracker Video Analysis and Modeling Tool to take accurate measurements based on the videos you'll make of the experiments. Tracker, which you can download from *opensourcephysics.org,* is free, open source software for Windows, Mac, and Linux.

Of course, it goes without saying that you will need access to a computer to take advantage of this tool. Another valuable software tool that will aid in your analysis is a spreadsheet program such as Microsoft Excel.

In every chapter of the book, you'll build a hands-on project. For these projects, I used kits and parts that were convenient for me to get. If you live in a different part of the country or a different part of the world, you may find other resources are more convenient that allow you to perform the same experiments. The step-by-step instructions I provide are guidelines. Like any good scientist, you should feel free to improvise!

We will start with a basic, but important type of motion. This is constant-speed motion. You'll build a constant-speed vehicle and think about what it means for an object to move at one speed. I'm ready to go, are you?

CHAPTER 1

Cruise Control: Constant Speed

PHYSICS IS THE MOST FUNDAMENTAL OF ALL the sciences. It is the study of the basic things the universe is made of—matter, energy, motion, and force. Physics tells you how these things work together. Let's begin your study of physics by building a project that you can then use to perform EXPERIMENTS. Experiments, measurement, and analysis are the foundations of physics and of all the sciences.

In this chapter, as in all the chapters that follow, you'll put together a do-it-yourself project that teaches you something about physics. After constructing the project, you'll use what you built to perform experiments. In this first chapter, you're going to learn several ways to take measurements, as well as how to analyze the DATA you get from those measurements. Scientists make sense of the world by creating experiments, taking measurements, analyzing data, formulating equations from the data, and, finally, coming up with conclusions. Scientists discuss their conclusions and experiments with others; in this way, the ideas of science are communicated throughout the world.

For the first project, you'll build a toy car that you'll then use to look at constant speed; you'll measure it, analyze it, and see what conclusions

you can draw. Motion is one of the fundamental things studied in physics.

Throughout this book, I also introduce you to people from history who made significant contributions to our understanding of physics. On the next page, you'll read about a famous scientist whose contributions I will refer to again and again.

Project: Build a Constant-Speed Vehicle

For this project, I take you through the steps used to build a model battery-powered vehicle from a kit. (See the "Resources" appendix for more information on the kit I used, along with other possibilities.) The kit I used didn't come with building instructions. To perform the experiments and measurements later in the chapter, you can build any model vehicle, as long as it moves at a constant speed. You can also build a constant-speed car from scratch rather than from a kit, but some of the parts needed may be hard to find at local hobby stores or hardware stores.

Things You'll Need

Parts
- Gear wheels (rear)
- Front wheels
- 3 volts (V) DC motor

 NOTE

DC means *direct current* and is the type of motor used with batteries. AC means *alternating current* and is the type of electricity found in most homes. For these projects, you will need to purchase DC motors.

- Package of four pulleys
- Package of four gears
- Four rubber bands
- Switch
- AA battery holder (for 2 AA batteries)
- Axle rods
- Four screw eyes
- Wood base

Tools
- Ruler
- Pencil or pen
- Pliers or long-nose pliers
- Hot glue gun with hot glue sticks
- Wire strippers
- Soldering iron (optional)
- Drill with small drill bits (optional)
- Small nail and hammer (optional)

 **BE CAREFUL!**

Hot glue guns can get *very* hot and cause serious burns. The electric drill can be dangerous, too. Always have adult supervision when you use any tool that can burn or cut you.

Famous Scientists

Galileo Galilei (1564–1642) was a famous physicist and astronomer who lived in Italy. Galileo's writings form the basis of KINEMATICS (kin-*uh*-mat-iks), which is a physics term for the study of how everyday objects move. You'll use many of Galileo's ideas throughout this book.

Galileo built telescopes, turning them to the night sky to look at the moon, planets, and the stars (astronomy). He also studied the motion of the moon and the other planets. Shown here is one of his drawings of the phases of the moon. One of Galileo's goals was to prove that all the planets, including Earth, revolve around the sun. At that time, many people believed that the sun and the planets moved around the Earth, which stood still. Galileo researched many things including constant speed and acceleration, projectile motion, and how objects move when falling toward Earth.

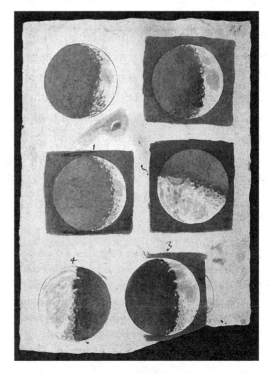

CHAPTER 1 ■ CRUISE CONTROL: CONSTANT SPEED

FIGURE 1-1 Snapshot of my completed constant-speed vehicle

Assemble the Kit

These steps describe one approach to assembling the Kelvin CV model racer shown in Figure 1-1.

1. Spread the components on a clean, flat surface and put similar items together.

2. Use the eyehooks to attach the axles to the car's body. Using the ruler, first make four marks on the wood base with a pencil. Measure ½ inch from the ends along both lengths, and ¼ inch from the ends along both sides of the width of the rectangular wood base (Figure 1-2).

3. Using these four measurements, draw four lines, with darker dots at the intersections.

These measurements don't have to be exactly like what's shown in Figure 1-2, but be sure to keep your lines straight.

4. Drill four small, shallow pilot holes for the eyehooks at the four dots you drew in step 2.

5. Screw the four eyehooks into the wood base. Rather than using a drill, you could also use a small nail and a hammer to create small pilot holes. You may need pliers to screw in the eyehooks securely. After the eyehooks are screwed in and aligned, make sure that your two axle rods fit into the eyes of the eyehook. (Look ahead at Figure 1-4 for a preview of the finished wheelbase.)

6. Once you have the axles lined up, remove them and insert one small front wheel on one of the axles and one of the larger wheels on the other axle. The small wheel is for the front of the car and the larger wheel is for the back of the car. The wheels should push onto the axles, although you may need to gently tap them in with a hammer.

7. Taking one of the plastic pulleys that came with the kit, insert the second-largest pulley onto the back axle so it is pushed close to the rear wheel; see Figure 1-3.

8. Push the axles through the eyehooks, and place the second front wheel onto the front axle and the second rear wheel onto the rear axle (Figure 1-4).

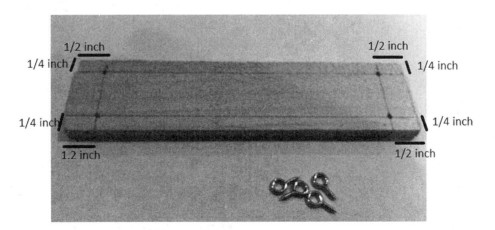

FIGURE 1-2 Creating the four marks with a pencil

FIGURE 1-3 Placing the pulley on the back axle

9. Once you finish the wheel assembly, push the car around to make sure it rolls in a straight line and with minor friction. Now you're ready to work on the motor assembly.

FIGURE 1-4 The completed wheel and axle assembly. The rear wheels are larger and the pulley is part of the real axle assembly.

10. Find the battery holder, switch, and motor. By making a complete circuit, these three devices make up the electrical and power-generation parts of your car.

11. There are four wires: two coming from the battery pack and two coming from the switch. Before you connect any of these wires, strip more from the ends using a wire-stripper so you have about ½ inch of stripped wire from each end.

12. Put two fresh AA batteries into the battery holder, and make sure the batteries are aligned correctly: positive (+) to + and minus (−) to −.

13. Take the red wire from the battery holder and twist it with one of the wires from the switch (see Figure 1-5). One option is to use the solder and a solder gun to make your connections more permanent and durable. I found that by twisting my wires together tightly, soldering wasn't necessary.

🧪 BE CAREFUL!

Burns from a solder gun can be very painful and severe. If soldering, do it under adult supervision.

FIGURE 1-5 Making a connection from the battery holder to the switch

14. Notice that the motor has two connections (A). Take the other wire from the switch and run it through one of these connections (B). Make sure to run the wire through the connection tightly and then twist the wire around the connection to secure it.

make sure your batteries are properly and securely placed in the battery pack.

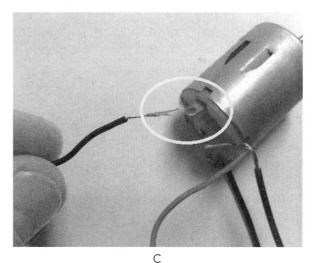

C

A

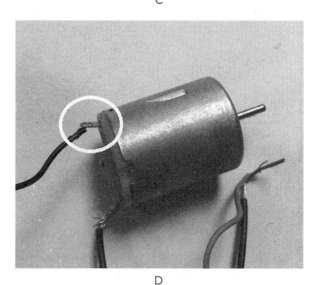

D

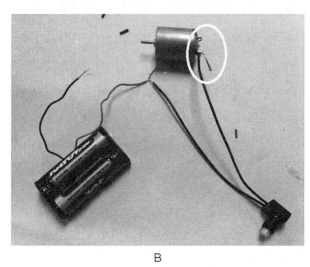

B

15. Now you're ready to make the final electrical connection. Thread the other wire from the battery pack through the other motor connection (C). Again, tightly twist it so you have an electrically sound connection (D).

16. Turn on the switch and see the motor's axle rotate. You have created a pathway for the electrical energy to run from the batteries through the switch to the motor. If the motor doesn't work, first check that your connections are making good electrical contact, and then

17. Put the small pulley on the motor's axle shaft.

FIGURE 1-6 Attaching the rubber band to the pulleys

18. You are now ready to add the rubber band to the pulleys on the motor shaft and the rear axle. Put the battery pack, switch, and motor on top of the car's wood body toward the middle and front of the car (see Figure 1-6).

19. Put one of the rubber bands around the motor shaft pulley and then around the rear wheel and around the pulley on the rear axle. There should be some tension or tightness in the rubber band, like Figure 1-7, but not excessive tightness. If the rubber band is too loose, it will fall off the pulleys too easily. If the rubber band is too tight, there will be excessive pull against the motor.

20. Once you have the battery pack and motor in position, hot glue the motor, battery pack, and switch to the wood top of the car.

BE CAREFUL!

Remember, hot glue guns can cause serious burns.

21. Retest your car's switch and electrical connections. Make sure pushing the switch activates the motor and spins the motor shaft, turning the rubber band and the back axle. You may need to make some minor adjustments.

22. Put the car on a smooth floor and turn on the switch. The car should roll forward in a straight line. You may want to test the car on different

FIGURE 1-7 Positioning the motor and battery pack

surfaces. You might find your car moves better on carpet than on a smooth floor. If your car is not moving in a straight line, check the straightness of your axles; you might need to make minor adjustments to the eyehooks. Make sure the wheels have adequate spacing and are not rubbing up against the side of the car.

Figure 1-8 shows two views of your new car.

Experiment
Determine the Speed of Your Vehicle with a Stopwatch and Meter Stick

For your first experiment you need

- Your constant-velocity vehicle
- Stopwatch
- Meter stick
- Smooth surface
- Software to graph your data

Physics is a science, and science involves learning things through experiments. In an experiment, you

A   B

FIGURE 1-8 Two views (A and B) of the constant-speed car

test a question or a HYPOTHESIS. Will your little vehicle move at a constant speed? That's what you want to determine. In this experiment, you

1. Acquire data.
2. Graph the data to construct a distance-time graph for the vehicle.
3. Determine the speed of the vehicle by taking the slope of the line formed.
4. Draw CONCLUSIONS concerning your question or hypothesis.

You can complete this experiment in several ways. Probably the easiest way is to run your car in a long hallway or room—like you might find at school—or a flat, smooth surface like a driveway, and use a stopwatch to measure your time. Either way, your testing area should be around 4 or 5 meters (around 5 yards) long. You need to measure the course and have a stopwatch handy.

 **NOTE**

This book uses the metric system (centimeters, meters, kilometers) rather than the *United States customary units* (inches, feet, miles) because the metric system is more commonly used in scientific studies. See the "Resources" appendix for some tips on conversions.

Using a meter stick, create a 4-meter (m) grid marked off in 1-meter units (at 0 m, 1m, 2m, 3m, and 4m), as shown in Figure 1-9. Start the car a short DISTANCE (say, half a meter) before the starting line to allow the car to get "up to speed." (Even though this is a constant-speed vehicle, the car needs some time to get up to its cruising speed.)

Start the stopwatch when the car gets to the beginning of the track and then time how long it takes to go 1 meter. After this, return the car to the starting position and time how long it takes the car to go 2 meters, then 3 meters, and finally 4 meters.

I timed each distance for two TRIALS and then took an AVERAGE time for each distance.

FIGURE 1-9 A test run for the model car

It's important to repeat things several times in experiments—that's the definition of a "trial." Repetition lessens the chance of any errors and makes for a better set of data.

To find an average, sum, or add, the data and then divide by the number of trials performed.

You can use a data table like the one I used here for your experiment:

Distance (meters)	Time (in seconds), Trial 1	Time (sec), Trial 2	Average Time (sec)
0	0	0	0
1	1.46	1.6	1.53
2	2.71	2.81	2.76
3	3.63	4.03	3.83
4	5.01	5.39	5.2

NOTE

Your data will most likely differ from mine because your car may move at a different speed.

Graph Your Data

You can turn your raw data into a distance-time graph using a ruler and graph paper. I created a graph using a spreadsheet program called Microsoft Excel, which can be faster and easier. You probably have access to this software on your school computers or at home. If you're not sure, ask a teacher or parent to show you how to use the program. Other spreadsheet programs work just as well (see the Resources appendix for more information), and many of them are free and available online.

NOTE

Spreadsheet programs are a popular tool for scientists because they allow you to do multiple calculations at once and to substitute new data for old easily. A calculator serves the same purpose, but with a calculator, the operations must be entered by hand each time. A spreadsheet is more efficient!

If you have compiled a grid of your own data from the trials you ran, you can transfer it to a spreadsheet program and make your own graph.

Using Excel, I created a DISTANCE-TIME GRAPH by combining both columns, averaging time and distance, and instructing Excel to place the average time on the *x* axis and the distance on the *y* axis. You can see my graph in Figure 1-10.

This graph makes a diagonal line called a TRENDLINE. A diagonal line on the distance-time graph means my car is traveling at a constant speed. (In the next chapter, you'll find out that other shapes on a distance-time graph mean other types of motion.) This is important as it tells me the type of motion.

If you're using Excel, once you've created a graph, you can right-click on one of your data points. Then choose **Add Trendline** from the menu.

Now you see a screen with Trendline options. Select **Linear**. Click to check the two boxes toward the bottom: **Set Intercept = 0** and **Display Equation on chart**. For 0 distance, your time is 0 seconds. This is your *y* intercept, and it should be 0. Consequently, your graph should go through the ORIGIN, and checking the **Set intercept = 0** box forces the graph through the origin. Close the window.

NOTE

The origin of a graph is typically the location on a graph where the two graphing axes meet and is labeled (0,0). For your purposes, this signifies that at *zero* time your car has moved *zero* distance.

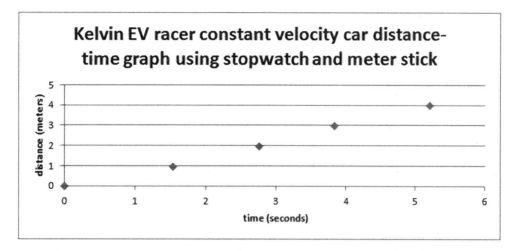

FIGURE 1-10 Excel graph based on my stopwatch and meter-stick data

Excel shows your trendline and EQUATION, so you can see your SLOPE. This slope represents the SPEED of the car (Figure 1-11). The *y* represents the distance, *x* is the time, and the slope, represented by *m,* is the speed of your car.

The slope for my vehicle is 0.7614 meters per second. Speed is a distance measurement divided by a time measurement. If you measure your grid in meters and time the car's motion in seconds, your speed is shown in "meters per second."

My car's slope or speed is slightly greater than 0.75 meters per second or 75 centimeters per second. Each second the car moves approximately 75 centimeters. Your car will probably have a different speed depending upon its motor, batteries, and weight. Regardless of your car's speed, its graph should be diagonal and represent constant-speed motion.

This equation, known as a kinematics equation, is called a *constant-speed* (or *velocity*) equation.

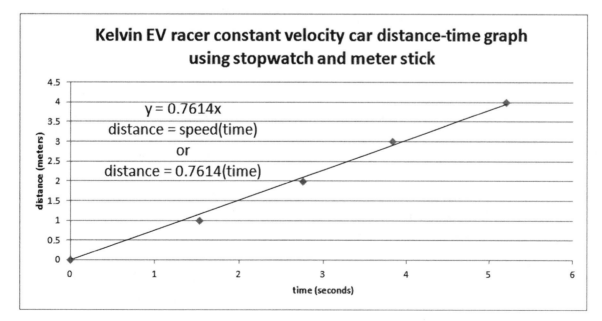

FIGURE 1-11 My car's speed and equation

It works for any object traveling at a constant speed. The object could be a car, plane, train, planet, satellite, or galaxy. In physics, we say that distance traveled equals speed multiplied by traveling time. In concept form, you would write:

Distance traveled equals *speed* multiplied by *time traveled*

Or you can identify speed as distance traveled divided by time. Here is how to look at it in the form of an equation:

$$Speed = \frac{Distance}{Time} \text{ (speed is defined as distance per time)}$$

This equation is powerful. By knowing and using this equation, you can make PREDICTIONS. How far would the car have moved in 10 seconds? Take your speed and multiply it by 10 seconds.

Using my data, 0.7614 multiplied by 10 seconds gives a distance of 7.6 meters. The car would move a little more than 7.5 meters if you let it travel for 10 seconds.

You can use this equation for a real car, too. If you set the cruise control so the car travels along the highway at a constant speed of 60 miles per hour (96.5 km/hr), then you can use the constant-speed equation to predict how far the car will have traveled in three hours.

 **NOTE**

Cruise control is an instrument in a real car that forces the car to remain at a constant speed. Ask an adult to show you how this instrument works.

Physicists like to find equations because equations help make predictions. And we often discover an equation by graphing data and looking at the shape of the graph. The data graphed comes from the experiments we perform. Although you can certainly graph the data with graph paper and pencil, computer software tools make it easier and faster to analyze the data and see the trendlines.

Experiment
Use a Video to Determine the Speed of Your Vehicle

Now I'll show you a second technique for taking measurements—one that is very popular in physics classrooms and involves using computer software.

For this, you need

- Digital video camera (or a phone or camera that takes video)
- Meter stick
- Tracker software (see "Resources" appendix)

With a stopwatch and a meter stick, you can obtain time and distance data that you can then use to graph the data and find the slope or speed of your object. In some situations, however, you might have trouble using a stopwatch and a meter stick. For instance, the object may be traveling too fast to measure easily with a meter stick. Another method of taking measurements involves taking a short video of the object in motion. You then analyze this video using motion analysis software.

Take Video of Your Constant-Speed Vehicle

You can use a cell phone, a video camera, or a digital camera set to movie mode to make the video. Any of these will work as long as you are able to transfer the video to a computer so you can analyze it later. You need to include a meter stick or ruler or some object with a known length in your movie. This is called a *calibration stick*, and it gives the motion analysis software a frame of reference to use as the constant-speed vehicle moves in profile in front of the stick.

Set your camera far enough back from the subject until you get a good field of view that covers the entire range of the meter stick. I like to use a small tripod to keep the camera steady; however, you can use duct tape to hold your camera in position on a table or counter or prop it against a book. The key is that the camera not move during filming, or video analysis won't work. Make sure the area where you take the video is well lit.

Digital video cameras take video at a speed of 30 frames per second, which is the frame rate you need to process your videos with the Tracker Video Analysis software. Most cameras store video in AVI or MOV format and the software can process either. Because digital video files are 30 frames per second, the software determines the time that the object is in motion. By including a meter stick or *calibration stick* in the movie file, the software then determines the distance moved. Using distance and time, the software creates data and graphs and displays how the object moves.

NOTE

Because of the advantages of taking videos to determine the nature of motion for many physics experiments, you will use this technique in many places in this book. I've placed all of my videos on the associated website for this book so that, even if you don't film your own video, you can still view mine and examine the analysis.

Set Up Tracker Video Analysis

Tracker is a free video analysis and modeling software package created by Douglas Brown and designed for physics education. Use your search engine to locate the Tracker Video Analysis and Modeling Tool online, or refer to the "Resources" appendix section. By downloading Tracker onto your computer, you will have a powerful program to analyze not only the constant-speed vehicle video but also other videos you will film in other chapters of this book. Tracker is easy to use with both AVI- and MOV-formatted movies.

NOTE

Tracker is a free tool for Windows, Mac, and Linux. You can download it from a number of sites, but you can find the OSP Tracker download page at http://www .opensourcephysics.org/items/detail .cfm?ID=7365.

HINT!

Save your Tracker file at regular intervals so you don't risk losing any data.

1. Open the Tracker software, and then open your video file in Tracker by selecting **File | Open**.

2. Click the **Calibration** button and select **Calibration Stick**.

Calibration button

3. After selecting **Calibration Stick**, you'll see a blue **100.0**, which stands for 100 cm, and blue calibration stick on your screen. Right-click on the calibration stick to change the color of this line, if needed, to make it easier to see.

4. Move the 100.0-cm line so it spans the length of the meter stick, by dragging the left side of the calibration stick to the left end of the meter stick, and the right side of the calibration stick to the right side of the meter stick, as shown in Figure 1-12. The meter stick is 100-cm long. If you use a different length of meter stick, change the 100 to match the length of the meter stick you used. For example, maybe a 25-cm floor tile shows up in your video frames. You can use this as your "meter stick," but you need to change the 100 to 25 and shorten the line to match the width of the tile.

5. After adding the calibration stick to your video file, decide which frames to use. Use the buttons at the bottom of the window to play the video file and move it forward a frame and back a frame. In Figure 1-13, I've moved my file to frame 10, which is where I will start. Select the starting frame for your video.

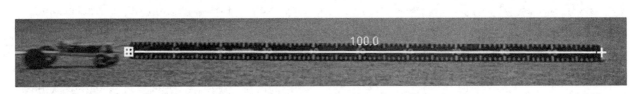

FIGURE 1-12 Drag the calibration stick so it's the same length as the meter stick.

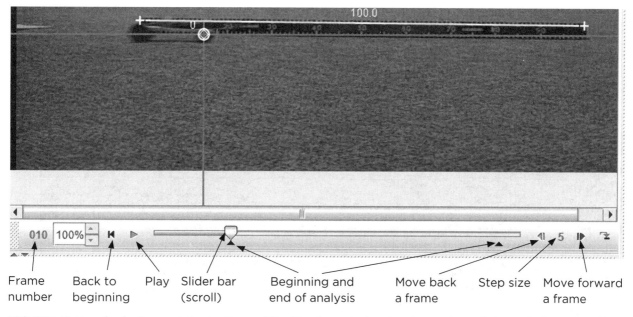

| Frame number | Back to beginning | Play | Slider bar (scroll) | Beginning and end of analysis | Move back a frame | Step size | Move forward a frame |

FIGURE 1-13 Use the buttons at the bottom of the Tracker window to play and scroll through the video file.

6. Then I moved to the point where the car is about to finish the run, frame 45 for my video file. Select the ending point for your video.

7. Now you need to put these frames into the instructions for the software. Click the **Clip**

Settings button to the left of the **Calibration** button at the top of the window.

8. The Clip Settings dialog appears, as you see in Figure 1-14, where you enter the beginning and ending frame numbers for your video.

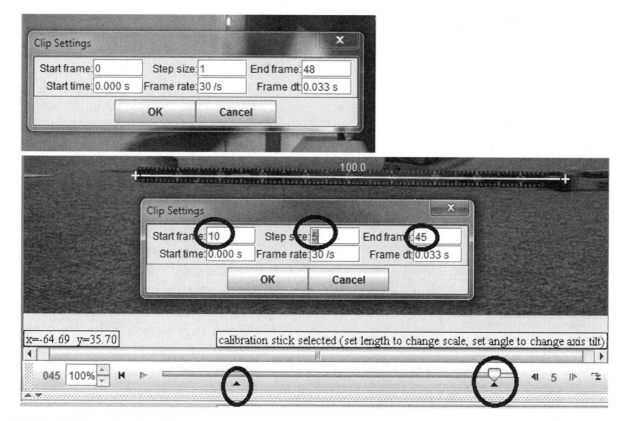

FIGURE 1-14 Changing the Clip Settings for the video file

For my video file, I started at frame 10 and went to frame 45. The **Step Size** represents the number of frames per step. Starting at frame 10 and ending at frame 45, I selected a step size of 5. Select the step size for your video. For this experiment, enter **5**. Using my video as an example, you "frame out" or use frames 10, 15, 20, 25, 30, 35, 40, and 45. This represents 8 total data points. I'll discuss how to "frame-out" these frames in step 10.

9. Move the slider bar at the bottom of the Tracker window to the left triangle, which is your starting frame. The car is now in the position shown in the picture below.

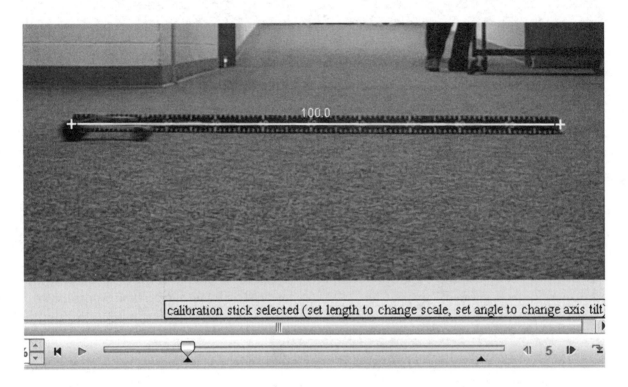

10. Now you frame out your video. To frame out your video, place a point on the same location on the object as it moves through the video frames. Click the **Create** button and select **Point Mass**.

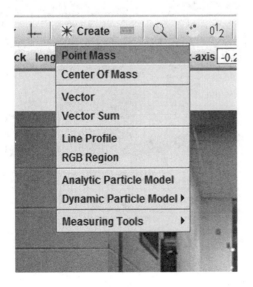

11. Move your cursor to the car's image and hold the SHIFT key. Notice the arrow cursor becomes a small box. Put the small box on a region of the car that you can see in each frame. I like to use the front or back wheel, and I like to put the box right in the middle of the wheel.

12. Left-click the mouse, and notice three things. First, the Tracker software has put a small red diamond and 0 indicator on the position clicked. Second, the software automatically moves the video file 5 frames ahead to the 15th frame, a five-frame step size. Third, the scroll bar has moved forward, too.

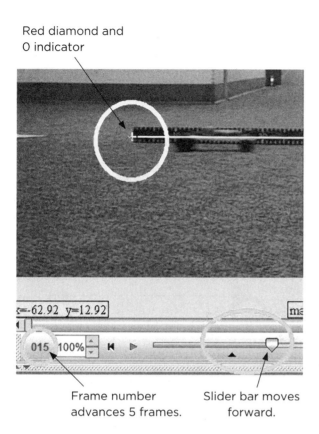

Red diamond and
0 indicator

x=-62.92 y=12.92

015 100%

Frame number
advances 5 frames.

Slider bar moves
forward.

13. Hold down the SHIFT key and make another mark on the same place on the car. You'll see another diamond and the number 1.

14. Continue to create these points until you reach the last frame. Now you've "framed out" out the video. The red color may be hard to see. When you place the cursor on one of the diamonds and right-click, you can change the color. Notice in Figure 1-15 that the diamonds are all about the same separation. This is one indicator that our car is moving at a constant speed or velocity.

If you picked a smaller step size, you would see more diamonds. For this car, five was a good frame size; the software has produced eight data points. If the car traveled faster, a smaller frame size would be a good choice. Notice on the right side of the screen, a data table and a graph was created as you set your point mass indicators on the car's rear wheel (look ahead at Figure 1-18 for an example). These numbers need to be linked to a correct coordinate system, so you have one more thing to do to correct your data table.

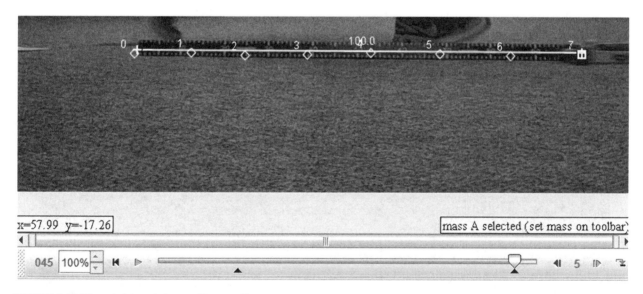

x=57.99 y=-17.26

mass A selected (set mass on toolbar)

045 100%   5

FIGURE 1-15 The eight points or diamonds

FIGURE 1-16 Set of coordinate axes in the middle of the screen

Click the **Coordinate Axes** button. A set of *xy* axes will appear in the middle of the screen, as shown in Figure 1-16.

The intersection of these lines represents the origin (0,0) point of your graph. Move the intersection to the 0 diamond so your first *x* data reading is 0 and your first *y* data reading is 0, too, as you see in Figures 1-17 and 1-18. The *t* data represent time, which also starts at 0. Move the cursor to the middle point of your axes, and the cursor changes from an arrow to an index finger. Move this middle intersection point until it is in the middle of your 0 diamond and your *x* and *y* values in your data table both read 0 and your graph starts at 0. This may take some maneuvering with your mouse. Remember to save your Tracker file!

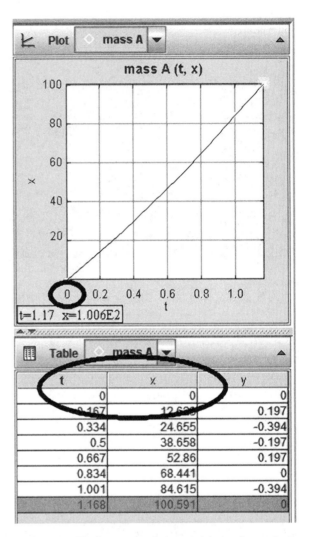

FIGURE 1-18 The corrected data table and graph

You may be wondering what the *x* and *y* mean. The *x* data represent forward motion. This is the data you want: the time (*t*) and the *x* data as you analyze the car's forward motion. The *y* data in the table represent vertical motion. For this experiment, the *y*

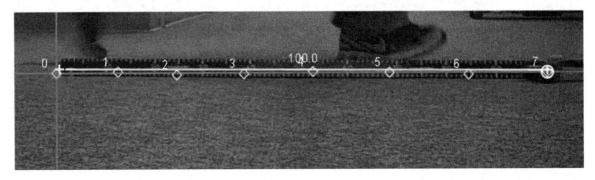

FIGURE 1-17 Setting up the origin correctly

is not important as the car is not moving vertically. In later chapters, I discuss the vertical motion of objects (a free fall, for example), and then, the y data becomes important.

The graph is displayed as a distance-time graph. Time is shown on the x axis and forward motion (distance) is shown on the y axis. The output graph is diagonal or linear. This represents constant velocity or speed. You knew this from the first experiment using the stopwatch and meter stick data. From that data, remember my car moved at a constant speed of 0.7614 meters per second (m/s).

On your Tracker menu, click **Window** and select **Data Tool (Analyze…)**. Check the **Fit** box. Some new options appear. **Fit Name** is a **Line**, and this option is correct as there is a diagonal linear relationship between the variables t and x, which represents the distance. The **Fit Equation** is **x = a × t + b**, with the a being the slope. Under Parameter (see Figure 1-19), a has a value of **86.318**. This is the slope in centimeters per second. My Kelvin EV racer is traveling at a constant speed of 86.318 centimeters per second (cm/s) or 0.86318 meters per sec (m/s).

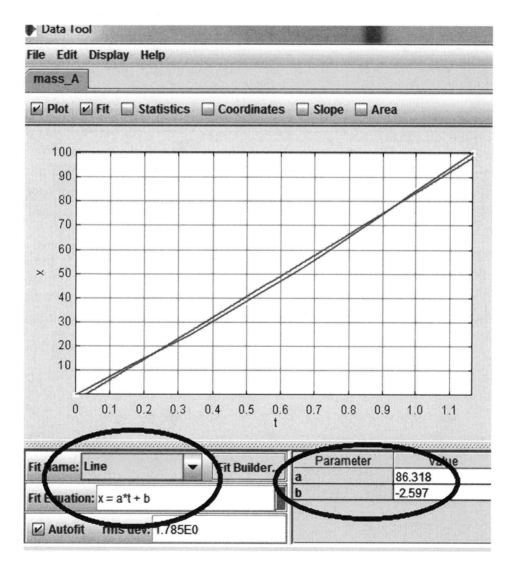

FIGURE 1-19 The car is traveling at a constant speed of 86.318 centimeters per second (cm/s), and there is a diagonal linear relationship between the variables t and x.

This is close to the value I got in the first experiment when I used a stopwatch and meter stick: 0.7614 meters per second (m/s). Both of your slopes or speeds should be similar numerically and within about 0.10 m/s of each other. In fact, both slopes should be the same. But the two methods you used to measure speed—the stopwatch and meter stick method and the video analysis method—use different techniques to extract the same information, the speed of the car. Each has potential sources of error and that is part of the business of doing science. But you have gotten close to the same speed using two independent experiments!

Summary

In this chapter, you learned about one type of motion—*constant speed*. You built a constant-speed vehicle, tested it, and measured its speed with several different experiments. You also learned how to use the Tracker software to analyze videos that you made of your car's motion. Constant speed, although the most basic and straightforward type of motion, is an important concept. Many types of objects on Earth and throughout the universe have unchanging motion, that is, the ability to move with constant speed. You can graph time and distance data for this type of motion. When you do this,

you get a diagonal line. The slope of this *trendline* represents the speed of the object. You can also use the *constant-speed equation* to make predictions:

$$Speed = \frac{Distance}{Time} \text{ (speed is defined as distance per time)}$$

I like to just take out my camera and video something or someone moving. Doing a quick analysis of my video tells me how fast the object or person is moving. Can you think of something around you that might be moving at a constant speed? Maybe you see a bug or ant moving along the sidewalk, or a friend running, biking, or skateboarding along a smooth level surface, a car driving along the road, or a plane flying across the sky. Pull out your smartphone or camera and take a quick video, and then analyze the data with a spreadsheet such as Excel or with the Tracker software. Graph the time on the x axis and the distance moved along the y axis. Do you get a sloped, diagonal line when you graph it on paper, with Excel, and with the Tracker software? If so, the object is moving at a constant speed, and you can determine its speed by calculating its slope and come up with its equation.

What about the motion of an object that changes speed? In physics, this is called *acceleration,* and this concept is the focus of the next chapter.

CHAPTER 2

Stop and Go: Changing Speed

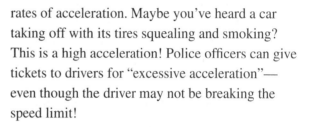

IN CHAPTER 1, YOU LEARNED ABOUT CONSTANT speed. Many objects move at constant speeds and understanding this type of motion is important for the type of motion you'll study in this chapter: CONSTANT ACCELERATION or UNIFORM ACCELERATION. ACCLERATION is a physics term meaning the rate at which an object speeds up or slows down over a period of time (with respect to time) is constant. If an object is getting faster, then it's called *positive acceleration.* If an object is slowing down rather than speeding up, it's called *negative acceleration.*

Constant acceleration means that the speed of an object is changing the same amount during each unit of time. Acceleration does not mean an object is moving fast, although it could be. An object could have a high rate of acceleration and still be moving slowly. Or it could be moving very fast and yet have little or no acceleration. Acceleration tells you how quickly speed is *changing,* not whether the speed itself is faster or slower.

Let's use an example from driving to explain acceleration. When a car is stopped at a red light, the car's speed is 0 miles per hour (mph). The car is not moving. When the light turns green, the car is ready to move through the intersection. The driver presses the gas pedal to move the car forward, giving it speed. This change in speed is called acceleration. The driver can press on the gas pedal with different pressures, which results in different rates of acceleration. Maybe you've heard a car taking off with its tires squealing and smoking? This is a high acceleration! Police officers can give tickets to drivers for "excessive acceleration"— even though the driver may not be breaking the speed limit!

Project: Build Your Fan Car

In the last chapter, you built a constant-speed vehicle. In this chapter, you'll build a small car that *accelerates* constantly, that is, a car that gets faster and faster. Small accelerating cars typically involve using batteries that spin a motor, which, in turn, rotates a propeller that pushes air. The pushed air propels the car forward at a faster and faster measurable rate. A car with a propeller is often called a *fan car*. A fan car accelerates for several seconds before it reaches its full operational speed

As you see in Figure 2-1, fan cars have many of the same components that you used to build the constant-speed car, including a motor, a battery pack, and a switch. As with the constant-speed car, the batteries provide the electrical energy that spins the motor. In this case, the spinning motor turns the propeller instead of directly turning the wheels. By installing a switch between the battery and the motor, you can turn the car off and on easily. You can make the car's body from just about any kind of material: balsa, straws, cardboard, and so on.

In this chapter, you'll build the fan car from scratch. For the do-it-yourself (DIY) fan car, you can get the parts you need from old toys or at electronic stores, hobby shops, hardware stores, or online. You can also build a fan car from a kit, which you can buy at a hobby shop or from an online retailer (see the "Resources" appendix). Either method works with the experiment in this chapter. If you would rather not build the car, you can still use the Tracker Video Analysis software and one of the videos included on the website for analysis.

This project is really fun, so let's get started!

FIGURE 2-1 Two different fan cars

Things You'll Need

Parts

- **DC motor (1.5 Volts to 9 Volts)**   Try to find a motor that is prewired (that is, with two wires coming from the motor already installed). Motors that are 1.5 V to 3 V run on one or two AA or AAA batteries. Each AA or AAA battery supplies 1.5 volts. *Voltage* is the push behind the electricity and the more voltage you have, the more your propeller spins. Both AA or AAA batteries supply the same push, but the AAA batteries are lighter weight. Make sure you get the correct battery pack for the size batteries you want to use.

- **Battery pack**   Try to find a prewired battery pack.

- **Batteries**   Get a fresh set for this project. You can use AA or AAA or 9 Volt batteries, depending on the DC motor you have.

- **Small DC switch**   Get a single pole, single throw (SPST) switch. This means the switch is either on or off. Different options will work, including toggle, push-button, and rocker switches. But don't get a "momentary" switch.

- **Insulated wire**   I used 18-gauge bell wire for my car, but 20 or 22 gauge would work too.

- **Four wheels and two axles**   The kind used for Cub Scouts pinewood derby cars work well.

- **Propeller**   Make sure your propeller fits the motor shaft. (Hobby stores geared to model airplane enthusiasts will have these.)

- **Car body**   Possibilities include cardboard, straws, foam board, and popsicle sticks. I bought a sheet of balsa from the hobby store (3/16" thick × 3" wide × 36" long).

- **Plastic straw**

Tools

- X-ACTO or hobby knife
- Ruler
- Scissors
- Tape (masking or scotch)
- Hot glue and a hot glue gun
- Wire strippers
- Needle-nose or long-nose pliers
- Solder and solder gun (optional)

BE CAREFUL!

Have an adult on hand to supervise this project. The X-ACTO or hobby knife is sharp. Cut over cardboard and be careful when using it. The hot glue gun can get very hot; be careful not to burn yourself. If you solder your electrical connections, remember the solder gun also gets extremely hot and can cause severe burns. And a rotating propeller can rotate fast enough to cut skin; take care around an activated propeller.

Ready to Roll? Steps to Build a Fan Car

1. Make sure you have a clean, flat surface on which to build. Take the material you're using for the car body and lay it on a piece of cardboard. For this example, let's use balsa. Using a ruler, measure a 4-inch length of balsa, as in Figure 2-2.

2. Cut the balsa (or other car-body material) along the line using an X-ACTO or hobby knife. Be careful with the hobby knife as you cut.

3. The width of the car is determined by the length of the axles. Measure the axles and cut the balsa based on their width. My balsa was 3 inches wide, so based on my axles, I cut the balsa to 1.5 inches. This is wide enough to support the battery pack and motor.

4. Use the plastic straw to create spacers for your axles. The spacers hold the axles and keep the

FIGURE 2-2 Put cardboard under your material to protect the table as you cut.

metal axles from rubbing against the car's base. The axles will rotate in the plastic straw, which decreases friction. A plastic straw of about any width works as long as the axle fits inside the straw easily. For my car, I cut two 1.5" pieces of straw.

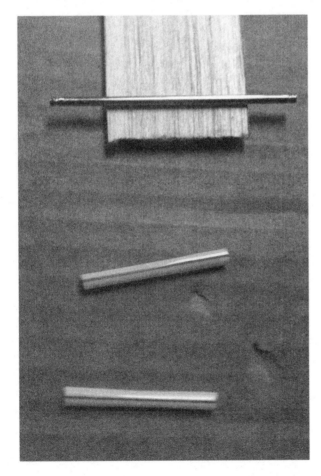

5. With a ruler measure ½-inch in from each end of the car's body to the end of the balsa body. Draw two measurement lines and place the two plastic spacers atop these lines, as shown here.

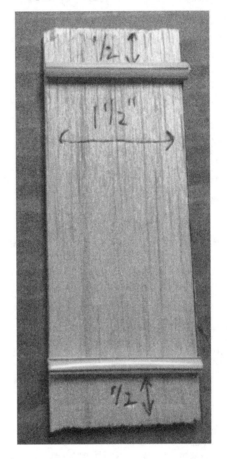

6. Use your hot glue gun to attach the straw spacers to the car's body. Line up the two spacers on the two lines you drew in step 5. Attach a small piece of masking tape to each spacer to temporarily hold it steady while you hot glue them both. Make sure the axles are straight, so the car rolls in a straight line.

BE CAREFUL!

Hot glue guns can burn; make sure an adult is present to supervise.

7. Slide the two axles through the straw spacers and then attach the wheels on the axles. Turn the car over and roll it around; it should roll smooth and straight with little friction.

The four wheels attached

Turn the car over and give it a roll!

8. Now you're ready to create the power circuit that will energize your motor. Gather the things you need: battery pack and batteries, motor, and switch.

 **NOTE**

If the motor and battery pack did not come prewired, you will also need insulated electrical wire to make the connections.

9. If your motor, battery pack, and/or switch came prewired, skip ahead to step 10. If not, you need to cut some wire and connect them. I used 18-gauge bell wire.

   a. Cut two pieces of wire using the wire strippers, each about 3 or 4 inches long.

   b. Strip about ½ inch of the insulation from the wire.

   c. The switch has two small connectors. One wire goes in each connector. Push the stripped part of the wire through the connector and then twist the wire around so it fits snugly around the connector, as you can see in Figure 2-3. You may need long-nose or needle-nose pliers to tighten your twist.

   If your motor needs wire, you need two additional pieces about the same length. Most battery cases come prewired, but if not, you need two wires for this, too.

10. Your motor, battery pack, and switch should each have two attached wires. Make sure the ends of each of the wires are stripped.

11. Take one of the wires from the battery pack and connect it to one of the wires from your switch. Twist the two stripped ends together until snug.

12. Connect the other wire from the switch to one of the wires from the motor. Again, twist the two stripped ends together.

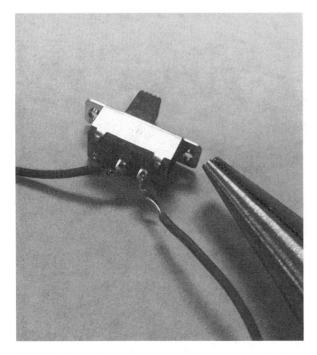

FIGURE 2-3 A close-up view of the switch with its two wires attached

13. Finally, connect the other wire from the motor to the other wire going to the battery pack. Twist these two stripped ends together, too.

14. Turn on the switch. The motor should now spin. If so, you have successfully connected the wires so they make a working electrical circuit. Make sure all wires are snug and that none of the stripped ends are touching other connections.

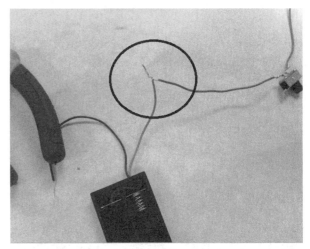

Step 11

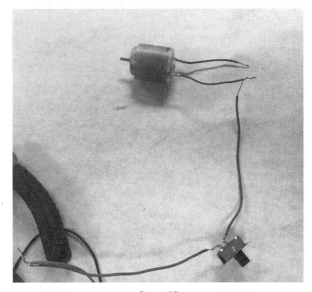

Step 12

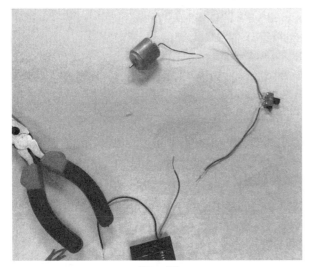

Step 10

Step 13

HINT!

Do not solder your connections yet!

15. Put the propeller on the motor shaft securely. It should push on tightly. If your propeller has a larger hole than the motor shaft, you may need to apply a bit of hot glue to secure the propeller onto your motor shaft.

16. Turn on the switch. You should feel air against your hand above your motor.

NOTE

If you feel more air behind the propeller, you may need to switch (or reverse) the wires coming out of your motor from your battery pack and switch. You want the connections that give you the most air flow against your hand above the propeller.

17. (Optional) Solder your electrical connections. If you've twisted your wires tightly and are making good electrical connections, you don't need to solder these connections. But soldering electrical connections can make for a more durable vehicle.

NOTE

Solder is an alloy of metals, usually tin and lead, and comes in rolls available at hardware stores. When heated by a soldering iron, the alloy melts and can be placed on an electrical connection. The alloy cools and solidifies again, joining the two wires together in a stronger electrical bond.

BE CAREFUL!

If you decide to solder your wires, be very careful and do so under adult supervision. A soldering iron can get hot, and burns from one can be extremely painful! Soldering is not difficult, but it does take practice. Always wear eye protection!

18. Make sure the rotating propeller clears the ground. If it doesn't, heighten the car base by adding several small pieces of balsa (or the material used for the car) to the end area of the base. Attach them with hot glue.

19. Hot glue your motor and battery pack to the base. Make sure the propeller is directed straight outward and not angled. For my car, I didn't glue down the switch but it's okay to glue it if you want to. Make sure that none of the wires impede your ability to change the batteries if needed or that they touch the ground as the fan car is moving.

HINT!

Make sure the propeller is glued on straight so the air flow is not directed at an angle.

20. Give your fan car a test run on a smooth floor to make sure the car rolls straight. If your axles are not perfectly straight, you can sometimes fix this by pushing gently on the straws until they are right. If your car is still not rolling straight, you may need to re-glue the wheels so they are straighter.

Experiment

Film the Fan Car's Motion and Analyze the Video

Building the fan car is a fun project, but let's do some science with it! In the first chapter, you used a stopwatch and meter stick to collect data. This approach won't be practical for the fan car as the acceleration phase of the car's motion occurs too quickly to be measured easily. But you can make a video of your fan car and then apply the Tracker Video Analysis software to analyze its acceleration.

Is your car accelerating as it gets up to cruising speed? If so, what is the value of this acceleration? How fast can your car go? These are all great questions that you can answer from analyzing a video. If you aren't able to make your own video, feel free to use the video I made of my own fan car, which is posted on the website for the book.

Set Up the Video for Analyzing

First, find a flat area for the car to run on. It doesn't have to be long; about 1 meter is long enough. Make sure you have space to move your camera back so you can video the moving car in profile. Use a meter stick (100 cm) or a yard stick (91.44 cm) to act as the calibration stick, and position the car so it does not run into the meter stick.

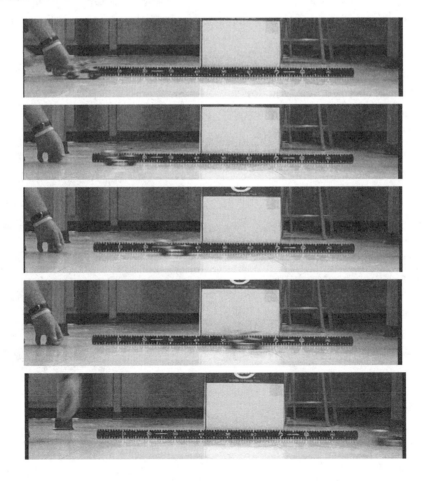

HINT!

Film in profile so you can see the motion of the car from the left end of the meter stick to the right end of the meter stick.

Take several videos. I always like to take more than one video just in case. Upload the videos to your computer. Now let's start analyzing!

1. Launch your Tracker Video Analysis software and open the correct video file.

2. Find the beginning frame in the video file to analyze using the slider bar and command buttons. You want to start the analysis at the frame when the fan car just leaves your hand (see Figure 2-4). This is frame 75 on my video file.

3. Find the ending frame in the video file to analyze using the slider bar and command buttons. You want to end the analysis when the car has moved by all or most of the meter stick. On my video file, this is frame 125.

4. Now you need to set up these frames for the software to analyze. Click the **Clip Settings** button to the left of the **Calibration** button at the top of the menu.

5. A prompt appears. Type in the beginning and ending frame numbers and enter a step size. Because I'm analyzing from frames 75 to 125, I set my step size at 5 frames. Move the video file back to your first frame, in this example, 75, using the slider bar button.

6. Create your calibration stick. Click the **Calibration** icon and select **Calibration Stick**.

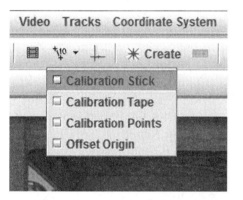

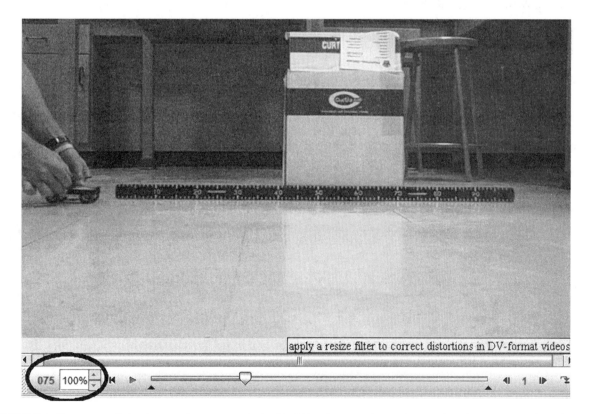

FIGURE 2-4 Use your slider bar and commands to find the frame where the car just leaves your hand.

7. Change the 100 (if needed) to match your calibration stick. Remember that this number is 100 centimeters. For my video, I changed it to 10 cm to match a 10-cm section of the meter stick. This step is important to get correct distances, speeds, and acceleration.

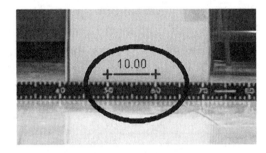

8. Now you're ready to frame out your video. Click **Create** from the menu bar and select **Point Mass**.

9. Move your cursor to the car's image and hold the SHIFT key. Notice the arrow cursor becomes a small box. Put the small box on a region of the car that you can see in all the frames. (I like to use the front or back wheel and I like to put the box right in the middle of the wheel.

I use the front wheel in my video file.) Left-click the mouse while pressing the SHIFT key. Notice three things:

- First, the Tracker software has put a small red diamond and **0** indicator on the position clicked.

- Second, the software automatically moves the video file 5 frames ahead to the 80th frame. This is the step size of 5 frames from step 5.

- Third, the scroll bar moves forward, too. Frame out your video file until you reach the last frame (frame 125 in my example).

Figure 2-5 shows the framed out video.

HINT!

Remember to save your Tracker file occasionally!

10. Click the **Coordinate Axes** button.

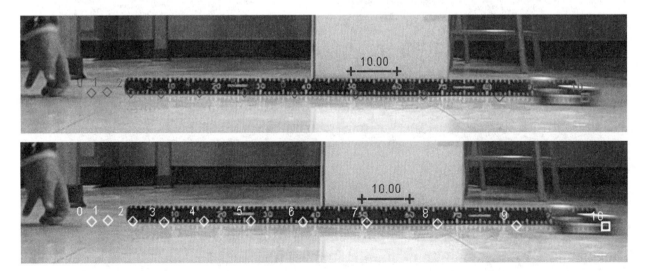

FIGURE 2-5 The top frame shows all the diamonds in place, and in the bottom frame, you can right-click one of the diamonds to change their color so they're easier to see.

11. You'll see a set of *xy* axes in the middle of the screen. The intersection of these lines represents the origin (0,0) point of your graph. When you move your cursor to the intersection of your axes, the cursor changes from an arrow to an index finger. Move this intersection point until it is in the middle of your 0 diamond and your *x* and *y* values are both 0, as shown in the photo here. This may take a bit of maneuvering with your mouse. Remember to save your file again.

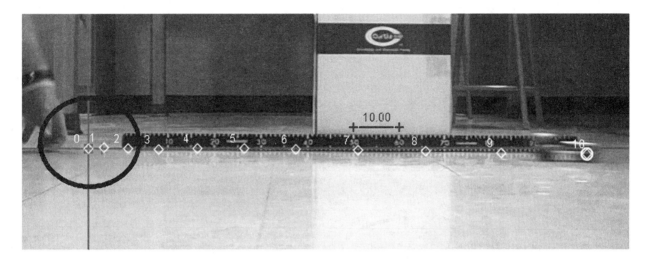

Determine Acceleration Rate with the Distance-Time Graph

Now that you have the fan car's video framed out, you can use the Data Tool to help analyze your car's motion. Doing this, you'll determine the car's acceleration and final operational speed.

1. Activate the Data Tool by selecting **Window** and then **Data Tool (Analyze...)**.

2. From the Data Tool window, click the **Fit** button and then select **Parabola** from the Fit Name drop-down menu, as shown in Figure 2-6.

Let's look at the graph above the data table in Figure 2-6. This is the same distance-time graph (*t, x* graph) you analyzed in Chapter 1. Look at its shape. Remember from Chapter 1, a straight line on the distance-time graph means the object moves the same amount of distance for each time interval. But now, the data points make a graph that curves upward. This curved shape is called a PARABOLA. In this case, the parabola is upward opening, so the fan car is accelerating positively— it's getting faster. This type of parabola means the fan car is moving an increasing amount of distance in successive time intervals.

Now let's figure out the rate of acceleration for the fan car. A parabola can be explained with three constants or numbers. They are shown under Parameter as *a, b,* and *c* in Figure 2-7. The important number here is the *a* value. For my fan car, I obtained an *a* constant of 29.404. I double this number to get my car's acceleration rate. If I double 29.404, I get 58.808 centimeters per second per second. The *a* value for your fan car will probably be a bit different.

🔍 INTERESTING FACT

Acceleration is a measure of how speed changes, or it is the double rate of a distance unit. If you use a meter stick for the calibration scale, then you measure acceleration as centimeters *per second per second*. You could write this as cm/s^2, but centimeters *per second per second* is simpler. Say a real car increases its speed by 10 mph per second; in this case, the units are miles/hour/second, so you have a distance unit divided by two time units. This is what I mean by *double rate*.

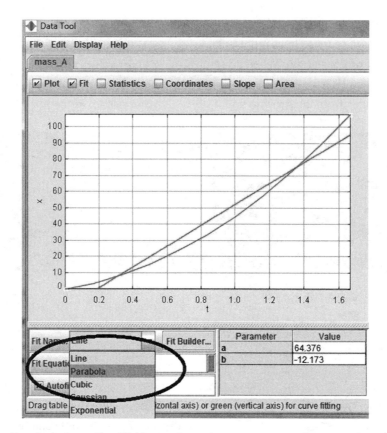

FIGURE 2-6 Selecting Parabola from the Fit Name drop-down menu

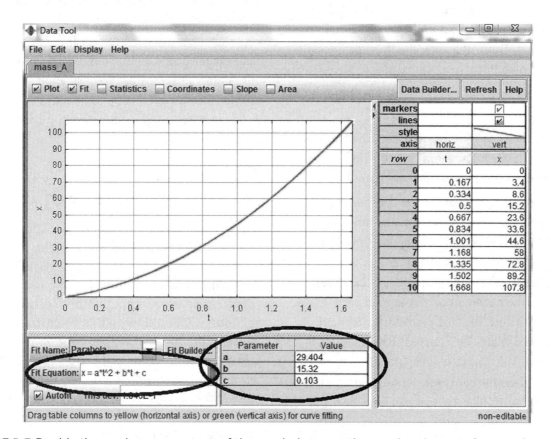

FIGURE 2-7 Double the *a* value, or constant, of the parabola to get the acceleration rate for your fan car.

Determine Acceleration Rate with the Speed-Time Graph

For an accelerating object, you get a parabola curved-shaped graph for your distance-time graph. You can use Tracker to create a second type of graph called a SPEED-TIME graph (also called a velocity-time graph).

1. Close the Data Tool window. You should see the (t, x) graph in the upper-right of the original Tracker window.

2. Click the *x* label to the left of the graph and choose **vx: velocity x-component** from the options that appear. A speed-time graph (t, v_x) takes the place of the distance-time graph (t, I).

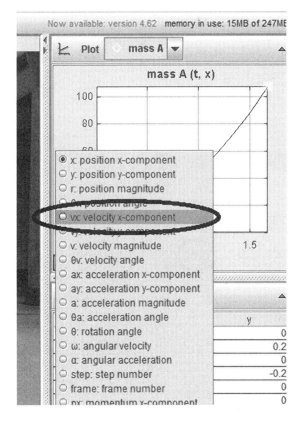

3. Return to the Data Tool window by selecting **Window | Data Tool (Analyze…)**. In the upper-right corner, circled in the next screen, uncheck the **Markers** and **Lines** options for the *x* row. The parabola will disappear from the graph.

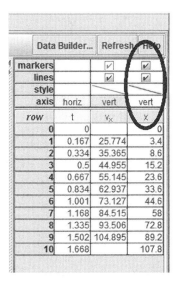

row	t	v_x	x
0	0		0
1	0.167	25.774	3.4
2	0.334	35.365	8.6
3	0.5	44.955	15.2
4	0.667	55.145	23.6
5	0.834	62.937	33.6
6	1.001	73.127	44.6
7	1.168	84.515	58
8	1.335	93.506	72.8
9	1.502	104.895	89.2
10	1.668		107.8

4. Click **Fit** and change the **Fit Name** to **Line**.

The speed-time data is linear or diagonal in shape. The speed-time graph in Figure 2-8 shows that the speed of my fan car is increasing by a regular value, or amount. You can also use the SLOPE of the line on the speed-time graph to get the acceleration fate for the fan car. What is the value of the acceleration? It is the slope of the speed-time graph, which is shown under Parameter *a*. For my fan car, the *a* value is 58.742 centimeters per second per second or 0.58742 meters per second per second. From this speed-time graph, you can derive an equation:

Change of speed of an object
= (its constant acceleration) × (time)

You can also use this graph to calculate your fan car's final speed. On the speed-time graph, look at where the line stops. In Figure 2-9, it is around 100 centimeters per second or 1 meter per second. If you look down at the time axis, this ending speed occurs around 1.5 seconds into the car's run. This represents my car's final speed and represents its maximum operational speed.

The fan car is a great project for learning about positive acceleration. The fan car starts at rest (0 speed) and its speed increases. Using the parabola on the distance-time graph, my fan car accelerated at a rate of 58.808 centimeters

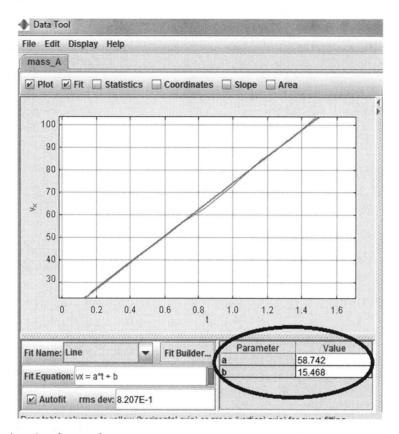

FIGURE 2-8 The acceleration for my fan car

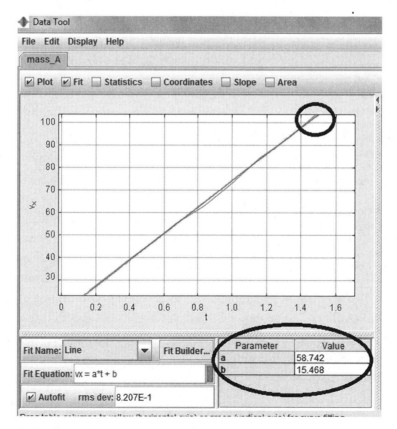

FIGURE 2-9 The fan car's final speed

per second per second. You can also use the slope on a speed-time graph to determine the acceleration. Using Tracker, I get a slope or acceleration of 58.742 centimeters per second per second.

Notice that you've learned two methods for determining acceleration. With one method (parabola on the distance-time graph), I get an acceleration of 58.808, and with the second method (the slope of the line on the speed-time graph), I get 58.742 centimeters per second per second. Both methods should give you numerically close rates.

Experiment
Use Tracker Video Analysis to Study Objects Slowing Down

Acceleration is changing speed with respect to time. What if an object slows down? What would the graph look like for this object? In physics, this is called *negative acceleration*. You can easily take an object, push it, and allow friction to slow it down until it stops. By taking a video of the object, you can use Tracker to see what distance-time and speed-time graphs look like for the object as its speed decreases.

I have included two videos on the book's website that show objects slowing down. But feel free to create your own video now that you have familiarized yourself with the tools. Take a toy car, give it a push, and let surface friction slow it down to a stop. Make sure to film in profile and include a meter stick or some object with known length. Make sure your object stops in your video.

 **NOTE**

If you need a refresher, turn back to "Set Up the Video for Analyzing."

Using the Tracker Video Analysis software, open the toy jeep movie (Toy Jeep Slowing Down) from the website and analyze it from frames 15 to 75 at a 5-frame step size, as you can see in Figure 2-10. In the video, I used a 100-centimeter meter stick as

a calibration stick. Use the Point Mass function to track the front wheel of the toy jeep and place the coordinate axes at the 0 diamond.

The first graph is the distance-time graph. In Figure 2-11, you see another curve shape, which is a parabola, too, but this time the parabola looks a little different. This parabola is opening more downward and suggests the toy jeep is moving a smaller distance during each successive time interval. Parameter a is now a negative number (−29.006), and when this is doubled, you get a negative acceleration rate of −58.012 centimeters per second per second. The negative number tells you the object is slowing down.

How does the speed-time graph change? Change your Tracker settings to see the (t, v_x) graph (you can follow the steps in "Determine Acceleration Rate with the Speed-Time Graph"). Return to the Data Tool window and unselect the Markers and Lines for x row. Change Fit Type to Line.

The slope of this line is the acceleration. But in this case, the diagonal is directed downward, and the slope is a negative number (see Figure 2-12). The negative sign tells you the toy jeep is slowing down. The Tracker software gives you an acceleration of −57.827 centimeters per second per second. Again, both acceleration rates are close.

For the second video, I pushed a weight horizontally along the floor and allowed surface friction to slow the weight down. Using Tracker, you can create distance-time and speed-time graphs for this video as well. Both the distance-time and speed-time graphs look similar to the jeep slowing down. You get a downward-opening parabola on the distance-time graph. The speed-time graph, which is shown in Figure 2-13, displays a negative slope that is equal to a negative acceleration. But look at the negative acceleration for the sliding weight as compared to the negative acceleration for the toy jeep. The toy jeep's acceleration was −57.827 centimeters per second per second, and the sliding weight's acceleration was −351.967 centimeters per second per second.

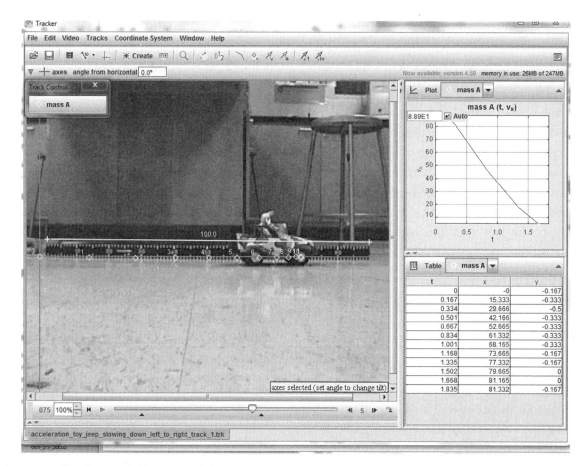

FIGURE 2-10 Tracker analysis video of the toy jeep slowing down

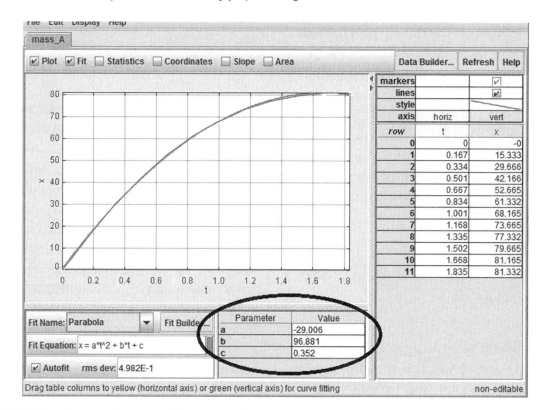

FIGURE 2-11 The distance-time graph for the toy jeep

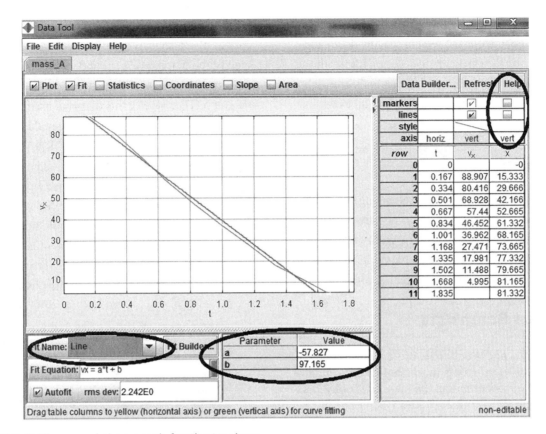

FIGURE 2-12 The speed-time graph for the toy jeep

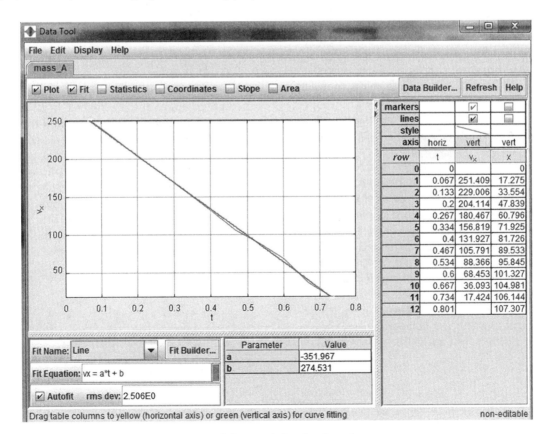

FIGURE 2-13 The speed-time graph for our sliding weight slowing down

The sliding weight slows down much faster! It takes about 0.7 seconds for the weight to slow down and stop, while it takes around 1.6 seconds for the toy jeep to slow down and stop! The weight's greater surface friction stops the weight pretty quickly.

Summary

In this chapter, you learned how changing speed can be either positive (speeding up) or negative (slowing down) acceleration. Distance-time graphs take the form of parabolas for objects accelerating. If a car gets faster, you get an upward-opening parabola, and if a car gets slower, you get a downward-opening parabola. You also learned about a new graph called a speed-time graph. For a car undergoing constant acceleration, you get a diagonal line on the speed-time graph. The slope of this diagonal is a measure of the acceleration. For a car getting faster, this diagonal is directed upward and acceleration is a positive number. For a car getting slower, this diagonal is directed downward and acceleration is negative. Knowing about acceleration is a key idea and building block of understanding motion. In the next chapter, you learn about a special kind of acceleration called *free fall*, and this concept, discovered by Galileo, is one of the key discoveries in physics.

Famous Scientists

Menaechmus and Apollonius of Perga were mathematicians and scientists in Ancient Greece, where many of the mathematical tools we use today in modern science were developed.

Menaechmus was born in what is now modern Turkey in the year 380 BCE. He was friends with Plato, the famous Greek philosopher. Although his writings no longer exist, Menaechmus was believed to have discovered the parabola and other conic sections. The conic sections are named because they represent shapes obtained when cutting a cone (conic) with a plane. In geometry, a plane is a flat two-dimensional surface.

Apollonius of Perga was an ancient Greek born in 262 BCE. In addition to studying mathematics, Apollonius was also an astronomer. It is believed that Apollonius coined the words used today for the conic sections: *ellipse, parabola,* and *hyperbola*.

Take a sheet of paper. If the sheet of paper enters the cone at such at angle that it remains in the cone, you have a *parabola* (number 1). If you could hold the sheet of paper horizontally and chop a cone in two, you would obtain a *circle* (number 2 in the illustration). If the sheet of paper chops the cone at an angle, entering and leaving the cone, you have an *ellipse* (number 2, top). Finally, if you have a cone above another cone and insert the paper vertically, you get a *hyperbola* (number 3).

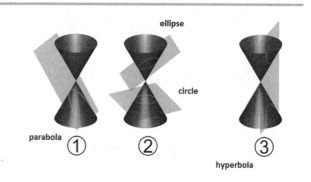

Parabolas and the other conic sections are important in many of the sciences. In physics, they are present on many graphs, including the parabolas on the distance-time graphs in this chapter. The orbits of the planets and other solar system objects can be described by the conic sections. Planets and satellites often travel in ellipses or circles as they orbit. Comets may travel in parabolas or hyperbolas. These types of orbits are often called open orbits as the object may make one pass around the sun and never return.

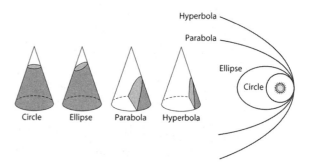

CHAPTER 3

Free Fall: What Goes Up, Must Come Down

UNTIL FAIRLY RECENTLY HUMANS COULDN'T escape the force of GRAVITY. Only in the last 50 years have we developed the technology to put satellites and spaceships into Earth orbit and beyond. Gravity is one of the fundamental forces of the universe. It is the attraction between objects that have mass. Our own bodies and the objects around us exert gravitational force, just as the Earth does. But because the Earth has so much more mass, we always fall toward it. The Earth's gravity pulls on everything, creating a force. How do you measure that force? What type of motion is occurring when you drop an object? Those are questions you'll explore in this chapter.

What Does It Mean for an Object to Fall Freely?

Galileo Galilei, who you met in Chapter 1, discovered that for objects moving at a constant rate of acceleration, you get a parabolic relationship when you graph distance against time (see Chapter 2). An object getting faster with a constant rate of acceleration creates a diagonal line on a speed-time graph, with a slope that is numerically equal to this acceleration. The distance-time graph for this object follows a parabolic shape. An object constantly getting faster travels farther and farther over each successive time interval.

Galileo reached these conclusions as a result of many different experiments involving rolling objects down long ramps and dropping objects from buildings. Legend has it that Galileo dropped balls and other objects from the tallest building in Pisa, Italy, his birthplace (see Figure 3-1). By dropping objects from the Leaning Tower of Pisa, Galileo is said to have concluded experimentally that all objects, regardless of weight, hit the ground at approximately the same time if dropped simultaneously from the same height.

Scientific thought up to this time was based heavily on the writings of Aristotle (384 BCE–322 BCE), the Greek natural scientist and philosopher. Aristotle had reasoned that objects fall based on their weight. By conducting experiments, Galileo discovered otherwise. He found that all objects undergo constant acceleration at the same rate when dropped, if dropped in the absence of AIR RESISTANCE. Galileo couldn't really remove the atmosphere and create a VACUUM in his laboratory, but he deduced that the air itself created a resistance that affected the rate at which objects fall. By performing experiments that lessened the effects of air resistance, Galileo was able to conclude that all objects undergo the same constant rate of acceleration when falling.

When objects fall due to gravity, it is called FREE FALL. Any frictional effects such as air resistance are ignored. Gravity causes the falling object to

FIGURE 3-1 Legend has it that Galileo dropped objects from the Leaning Tower of Pisa.

accelerate constantly. In this chapter, we will try to replicate Galileo's findings:

- All objects accelerate at the same amount if you ignore any frictional effects such as air resistance.

- The speed of a falling object increases as it falls, and you can see this increase as a diagonal line on the speed-time graph.

- The distance traveled by a fallen object becomes greater over successive time intervals, and you can see this as a parabola on the distance-time graph.

You will also discover the numerical value of this acceleration caused by gravity. Although Galileo's experiments led to these insights, it took later experimenters to discover the numerical value of the rate of acceleration caused by Earth's gravitational influence.

Famous Scientists

David Scott, NASA moon walker and moon driver, was part of the Apollo 15 mission to the moon. David Scott and his fellow astronauts, Alfred Worden and James Irwin, collected 170 pounds of lunar rocks, dirt, and dust, and performed many experiments. Scott also became the first astronaut to "drive" on the moon.

The hammer and feather experiment is one of the most famous scenes from the Apollo moon missions. In the photo shown here, mission commander David Scott is in front of the lunar lander holding a hammer in one hand and a feather in the other. In the movie, Scott drops both simultaneously. On the Earth, the hammer hits the ground first as the feather's motion is slowed by the atmosphere. On the moon, because there is no atmosphere but there is gravity, albeit less strong than Earth's gravity, both the hammer and the feather hit the lunar surface at the same time, just as Galileo predicted. The NASA movie can be found here—http://nssdc .gsfc.nasa.gov/planetary/lunar/apollo_15_feather_ drop.html—and it is definitely worth watching!

Experiment
Let's Drop Some Objects!

For this next activity, you need a couple sheets of notebook or plain white paper. Hold a flat sheet of paper in one hand and a crumpled sheet of paper in the other.

Drop them both at the same time from the same height. In this case, as you see in Figure 3-2, the crumpled sheet of paper hits the ground first. The non-crumpled sheet of paper is caught up by the effects of the atmosphere (*air resistance*), which slows its falling motion. Galileo hypothesized that if you could remove the atmosphere, both objects would hit the ground at the same time. Now take two pieces of paper and crumple them both together to form one ball. Drop the one-sheet-of-paper crumpled ball and the two-sheets-of-paper crumpled ball at the same time from the same height.

The two-sheets-of-paper ball is the one on the left in Figure 3-3. Both balls hit the ground at virtually the same time, even though the

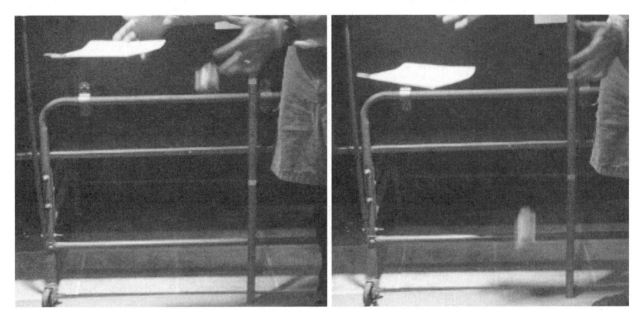

FIGURE 3-2 Drop a flat sheet of paper and crumpled sheet of paper at the same time. Which hits the ground first?

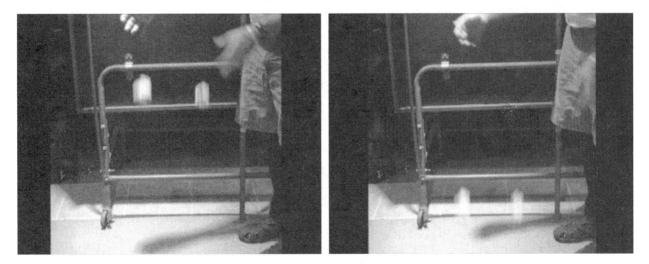

FIGURE 3-3 Drop two crumpled pieces of paper at the same time. Which lands first?

two-sheets-of-paper ball weighs twice that of the one-sheet-of-paper ball. As Galileo hypothesized, all objects, regardless of weight, fall at the same rate of acceleration if you ignore air resistance. Try dropping different objects. Make sure to pick objects that do not interact significantly with the atmosphere.

Now that you know for certain that all objects fall under the same influence of gravity, it's time to figure out the nature of this fall. Does the object fall at a constant rate of acceleration? If so, you should see a diagonal line on the object's speed-time graph and a parabola on the object's distance-time graph.

 HINT!

When performing this experiment, it's important to pick an object, such as a piece of paper, without much air resistance.

A Distance-Time Graph for a Falling Baseball

Measuring the rate of acceleration in free fall is an experiment best done with video technology. You used Tracker Video Analysis software in the first two chapters to analyze constant speed and acceleration. I filmed some videos of myself dropping various balls to include on the book's website, which you can also view. But I suggest creating your own video. Set up the camera on a rigid surface or tripod so you capture the complete fall and make sure to include a meter stick or other calibration stick in the field of view. Upload your video to your computer and use the following directions to analyze it. Your frame numbers will differ from mine, but the steps are identical.

I opened the Dropping A Baseball video in Tracker. Starting at frame 66, I framed the baseball as it fell through frame 78, at a step size of 2 (Figure 3-4). Instead of the time and *x* data, you need to view the

time and *y* data because the ball is falling in the *y* direction. Be careful here as your dropped object may blur due to its speed increasing as it falls. Be careful, too, not to include frames where the object is still in your hand or has hit the ground. Make sure to set up the calibration stick to match up with the meter stick (100 cm) and line up the coordinate axes to start with your first image (frame 66 on mine).

Because you're working with an object moving vertically, select the *y* data from the list in the Tracker software, as shown here.

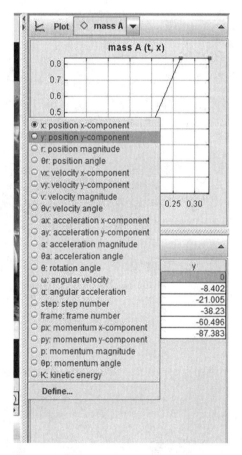

Now let's look at your distance-time graph. Select **Window | Data Tool (Analyze...)**. Check the **Fit** box and make sure to select **Parabola** for **Fit Name**. The *a*, *b*, and *c* numbers are the constants that define the motion of the baseball as it falls

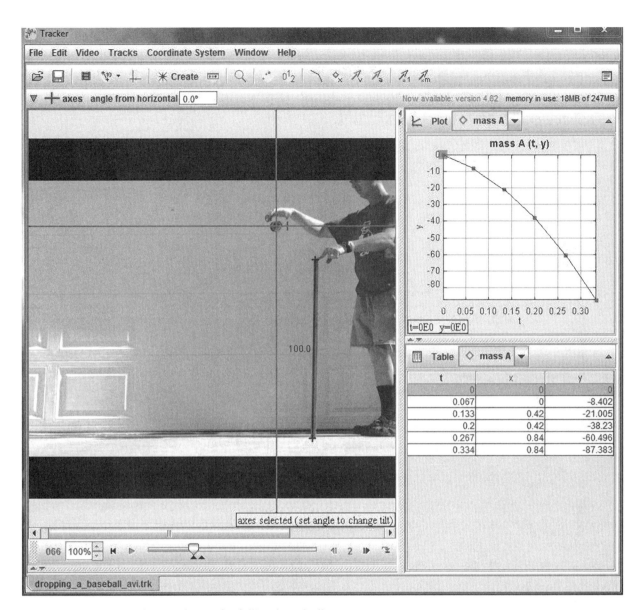

FIGURE 3-4 Tracker video analysis of a falling baseball

and are the constants for the parabola. Your screen should look similar to Figure 3-5.

This parabola looks a little different than the ones in the last chapter. Although it is a bottom-opening parabola—like the one you got as the fan car slowed down—you know the baseball is not slowing down! The baseball is getting faster as it moves in free fall. In Figure 3-5, you can also see that the baseball is traveling below the coordinate axes, and consequently, the distances are negative values. The negative signs do not imply negative numbers,

however, as these distances are truly getting larger. The negative signs instead imply the direction: the baseball is moving downward at increasingly greater distances during each successive time interval.

For the a parameter, my value is –525.585 centimeters per second per second. As in the previous chapter, you can double this value to obtain the acceleration of the baseball. When I double my a parameter, I get an acceleration rate of –1051 centimeters per second per second. You can convert this to –10.51 meters per second

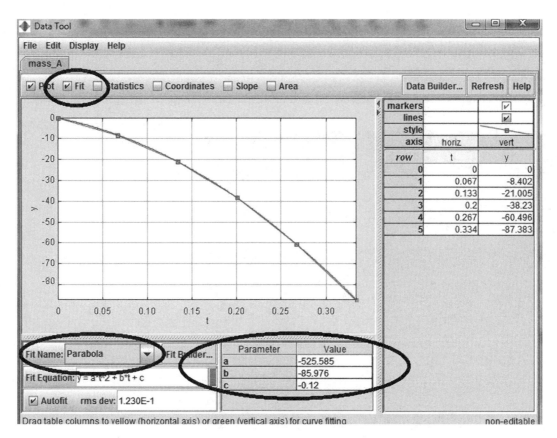

FIGURE 3-5 The distance-time graph for the falling baseball

per second. Again, the negative sign implies a direction; the acceleration is directed downward.

A Speed-Time Graph for a Falling Baseball

You can confirm this acceleration rate with a speed-time graph. First, you need to get the speed data from Tracker. To get the speed data, close the Data Tool window and then mouse to the *y* label on the graph. Click this to access the complete list of graphing options. Select **vy: velocity y-component**. Then select **Window | Data Tool (Analyze...)** again. You should see something similar to Figure 3-6 at this point.

To finish your speed-time graph, make sure to uncheck the **Markers** and **Lines** options above the **vert y** row at the upper right to delete the distance-time graph. Then select **Fit** and then select **Line** from the **Fit Name** drop-down list.

The slope of this diagonal is the numerical value of the constant acceleration and is the *a* under **Parameter** and **Value**. This number represents the effect of gravity on all objects that are falling freely. The slope of my graph is –1056.56 centimeters per second per second or –10.5656 meters per second per second. Notice in Figure 3-7 that the diagonal line is directed downward, giving me a negative slope. This negative sign represents a direction. Gravity pulls objects downward with a constant acceleration, and the baseball is getting faster as it moves downward.

Both of my graphs give me similar accelerations (–10.51 for the distance-time graph and –10.56 for my speed-time graph). More rigid experiments conducted by scientists give a value of approximately –9.8 meters per second per second for objects in free fall in a vacuum. In the United States, this rate of acceleration in free fall is often expressed as –32 feet per second per second. This is called

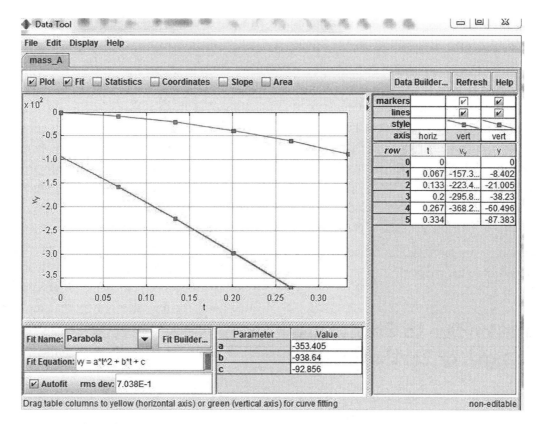

FIGURE 3-6 Data Tool window

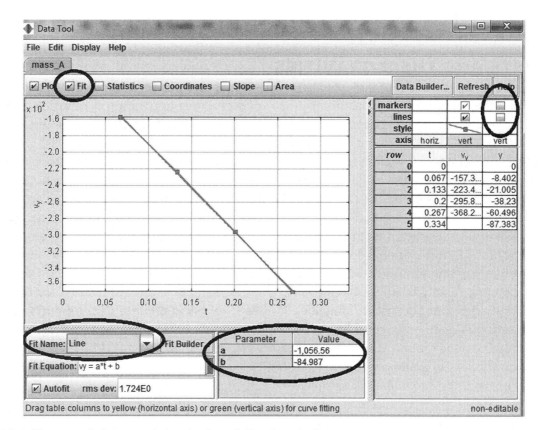

FIGURE 3-7 The speed-time graph for the free-falling baseball

the ACCELERATION DUE TO GRAVITY, and in physics equations, this value is given the symbol g (lowercase letter g).

My values of –10.51 and –10.56 are close to the accepted value of –9.8, though not right on. How about the values from your video? Are they close to –9.8? It turns out, whether you make 10 videos of different falling objects or 1000 videos, if you choose objects that interact very little with the atmosphere, you will get a rate of acceleration of approximately –9.8 meters per second per second.

Are your findings similar to Galileo's findings mentioned earlier in the chapter?

Acceleration Due to Gravity and the Strength of Earth's Gravity

Technically, the falling baseball isn't really in free fall because it is falling through the atmosphere.

But the effects of air resistance on the baseball are pretty small and your results probably show this to be true.

The constant rate of acceleration due to gravity for falling objects is equal or close to –9.8 meters per second per second. This specific number reflects the strength of Earth's gravity. Each planet has a unique number that specifies the strength of its gravity. Some planets have gravities much stronger than Earth and other planets have weaker gravities. For example, the acceleration due to gravity for Jupiter is approximately –30 meters per second per second. Its gravity is approximately three times as strong as on this planet. Mars, on the other hand, has a weaker gravity than the Earth's. Its acceleration due to gravity is –3.7 meters per second per second.

🔍 INTERESTING FACT

PLANETS AND THEIR GRAVITIES

Physicists use a lowercase g to indicate acceleration due to gravity. Each planet has its own gravity (g) that is based on the planet's DENSITY. Density is defined as the mass of the object divided by its volume. If you know or can estimate the radius and the mass of a planet, you can find the rate of acceleration due to gravity for that planet. For example, the acceleration due to gravity on the moon is –1.6 meters per second per second. When you analyze objects dropped on the moon, the resulting graphs have the same basic shapes that you get for objects dropped here on Earth. However, with a smaller rate of acceleration, an object dropped on the moon would need to fall farther to achieve the same speeds as seen on the Earth.

As mentioned earlier, the *average* rate of acceleration due to gravity (g) for Earth is

–9.8 meters per second per second. The value of g changes depending on latitude, depth, and the geological features of a certain location. Earth's rotation is not the same at all places on the planet and the planet is not a perfect sphere; both of these factors interact to change gravity a small amount. Earth's density also varies from place to place, depending on the type of rock beneath you. To help measure these small variations in Earth's gravity, NASA launched the GRACE mission in 2002. GRACE stands for *Gravity Recovery And Climate Experiment* and the mission involves a pair of satellites.

So how can two satellites in orbit, circling the Earth, measure these small variations of gravity? The two satellites are 137 miles apart in orbit. You can see them both in the photo. When the leading satellite encounters a region where the gravity is slightly stronger, it is pulled forward slightly farther than 137 miles. The distance between the two satellites can be measured very accurately. By measuring the distance between the two satellites, the small differences in gravity in different locations on Earth can be determined by satellite. Pretty nifty!

The lighter and darker areas on the map shown below represent areas where gravity is slightly weaker (you can see the full map online and in color at http://en.wikipedia.org/wiki/Gravity_Recovery_and_Climate_Experiment). Ocean trenches and depressions are lighter in this picture. Darker areas show places where the gravity is slightly stronger. For example, the Andes mountain range in South America is darker, a location where gravity is slightly stronger.

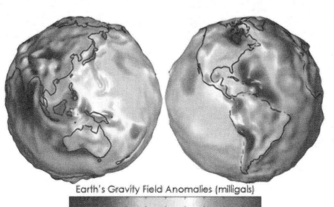

Earth's Gravity Field Anomalies (milligals)

-50 -40 -30 -20 -10 0 10 20 30 40 50

Project: Tennis Ball Cannon

It may seem counterintuitive, or even odd, but an object going up can be just as much in free fall as an object falling down. For instance, you can throw or kick a soccer ball in the air and measure the rate of negative acceleration as it climbs, reaches the APEX or top of the throw, and begins to fall. Gravity acts to pull the soccer ball downward, but because the soccer ball has upward motion due to your hands or feet, it takes time for gravity to slow the ball down. The ball slows down and eventually stops and then it returns to Earth. The soccer ball is coasting upward under the influence of gravity. It is in free fall but traveling upward. I filmed videos of such an experiment and put them on the website for this book. Take a look!

For the next project, you'll build a cannon that will shoot a tennis ball upward at a much faster rate than I can throw. But the results are the same. Just like the soccer ball, the tennis ball is shot upward and becomes an object in free fall even though it is moving upward. By taking a video of the tennis ball shot from the cannon, you can measure its speed as it moves upward under the influence of Earth's gravity. The project is a lot of fun!

BE CAREFUL!

The following project involves using flammable lighter fluid to create an explosion that hurls a tennis ball into the air. Anything involving flammable fluids, explosions, and projectiles is dangerous and should not be attempted without adult supervision.

Things You'll Need

Parts

- **6 tin cans (2 5/8" wide)**   Asparagus cans work great as they are the right height and width. Several brands of asparagus come in this size can. The cans must be this width and, they must be made of steel. You will remove both the top and bottom lids so make sure the bottom lid is flat and can be removed with a can opener.

- **1 tin can (2 5/8" wide)**   A tomato sauce can works well for this one. It must also be 2 5/8" wide and must also be made of steel.

The size of these cans must be exact. Take your ruler to the store!

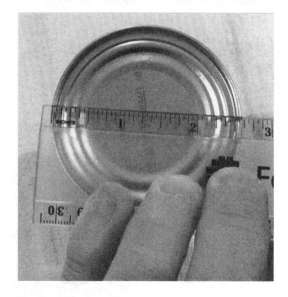

- Tennis ball
- Lighter fluid
- All-purpose long-arm butane lighter

Tools

- Ruler
- Roll of good quality duct tape (60 yards)
- Bottle opener (with triangular opener)
- Can opener
- Eye protection
- Ear protection
- Scissors
- Hair dryer
- Hammer and large nail (large enough to make an approximate 1/8" hole)

Let's start building the cannon:

1. Take the top lids off the six asparagus cans with the can opener. Save the asparagus to eat later—it's delicious and good for you! Clean out the cans, remove the labels, and remove the bottom lids with the can opener, too. You don't

need to save the lids. Your cleaned-out cans should look like this.

2. Take the duct tape and wind it several times around the seams of the six cans, attaching the cans together to create one long tube, as shown here.

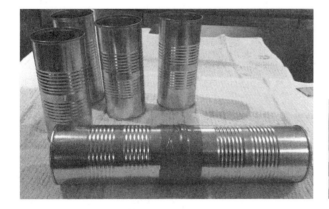

HINT!

Wrap the duct tape tightly around each seam several times. I worked with pairs of cans and then put them all together.

3. Now you're ready to construct the bottom of the cannon with the tomato sauce can. DO NOT open this can. Instead, use the triangular bottle opener and make six holes with the opener around the top of the can, as shown

here. Shake the tomato sauce out and save it for spaghetti sauce. Clean out the can.

 HINT!

I used a warm hair dryer to dry out the can.

4. With the hammer and large nail, drive a hole into the side of the can close to the bottom and then remove the nail. The hole should be approximately 1/8-inch in diameter.

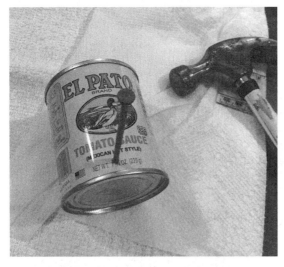

5. The tomato sauce can will be at the bottom of the cannon. The side of the can with the triangular holes is the top and should be duct taped to one end of the asparagus cylinder, as shown here.

Wind the duct tape around the cans several times and secure tightly. Here's a photo of what the completed cylinder will look like.

 HINT!

Doing this next step is easier with two people!

6. Wrap the rest of the duct tape roll around the cans to make a strong cylinder, as shown in

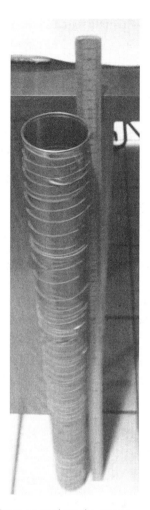

FIGURE 3-8 Your completed cannon

Figure 3-8. Starting from the top of the cannon work downward with the duct tape and then back up to the top. Do not cover the small nail hole at the bottom of the cannon (stop slightly above the hole). You want a strong tube that can withstand the explosion within the cannon. The more duct tape, the better! Once the cannon is securely taped, it will be ready to fire.

Fire the Tennis Ball Cannon

The tennis ball cannon is now ready to fire. You need a large, open, well-ventilated space away from houses. (It can't be too windy, either.) I wouldn't suggest indoor firing unless you have a large space with high ceilings with no windows or glass around

(I have fired the cannon in an empty school gym when it was too windy outside).

BE CAREFUL!

You will need eye and ear protection. Make sure an adult is present, too. Don't fire the cannon at other people and make sure any spectators are well away from your firing spot.

To fire your cannon, you need lighter fluid, a long-arm lighter, and a standard tennis ball.

1. With your eye and ear protection on, open the lighter fluid and either pour a little bit into the barrel of the cannon or into the nail hole. Either should work but you might find that one method is better than the other. I have launched my cannon using both fuel-injection methods.

HINT!

You don't need much fuel; maybe one squirt.

2. Grasp the cannon and wave it around in a sweeping motion for a few minutes. This moves the fuel around to all areas of the cannon and allows some of it to evaporate in the cannon. The cannon won't fire without doing this step.

3. If you plan on videotaping the launch, set up your camera away from the launch site but close enough to film the launch. I set up my camera a few meters away from the cannon.

4. Place the tennis ball in the end of the cannon. Don't push the ball too far down. Just push it down as far as you can with your fingers. The tennis ball should fit snugly.

5. Make sure your launch area is clear. Aim the cannon straight up and hold the long-arm lighter to the nail hole in the bottom can.

The flame from the lighter should ignite the fuel and cause the tennis ball to shoot out from the top of the cannon.

In my experience, waving the cannon tube after injecting the lighter fluid into the cannon tube is a key step. You may have to experiment with the amount of fuel to use. Don't get too hasty. I've injected the lighter fluid into my cannon, waved it around, and then set up my camera. After firing your cannon, check to make sure your duct tape is still secure and tight.

Experiment
Track the Tennis Ball

The tennis ball is shot out of the cannon so fast that, admittedly, it can be hard to get enough data points from a video. I did an analysis with both a vertical and a horizontal shot. I used my cannon itself as the calibration stick. Figures 3-9 and 3-10 show the Tracker screenshots for my vertical launch. (You can see the screenshots from my horizontal launch on the website.)

Open the Tracker software, load the video, and them frame it out, making sure to load the y data in your Data Tool window to create a distance-time graph. This analysis gives me a slope of around 2043.361 centimeters per second, or 20.43361 meters per second (around 45.6 miles per hour!). This is the approximate speed of the tennis ball as it is ejected from the cannon. Why approximate?

You know that once the tennis ball is shot out of the cannon, gravity is acting on it and the tennis ball will slow down. I projected a distance-time graph, and with only two points, you can see the two points as a diagonal in Figure 3-10, which implies constant speed. However, because gravity

FIGURE 3-9 Framing out the video for my vertical launch

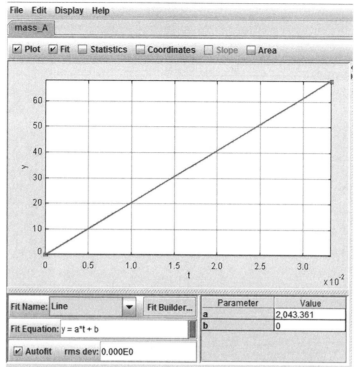

FIGURE 3-10 My vertical distance-time graph

is acting on it, you know the speed of the tennis ball is decreasing as it travels up. Because the time interval is small, this error is small.

Use a Physics Equation with the Tennis Ball Cannon

You can also use a stopwatch to determine the tennis ball's speed. Using a stopwatch, you (or maybe have a friend help) could estimate the time that it takes for the tennis ball to reach its maximum height. Then you use a physics equation:

$$0 = \text{initial speed} - 9.8\ (t)$$

This equation in its general form is

$$v_{final} = v_{initial} + at$$

At the apex or top of the trajectory, the final speed (v_{final}) is 0. Although there is some air resistance on the tennis ball, ignore this effect. The tennis ball is in free fall upward, so the acceleration (a) is –9.8 meters per second per second. Let's say that you shot off the tennis ball cannon and it took 2 seconds

for the tennis ball to reach the apex. You would have the following numbers for your equation:

$$0 = \text{initial speed} - 9.8\ (2)$$

Solving for the initial speed, you obtain 19.6 meters per second or around 44 miles per hour.

Summary

Free fall is a key type of motion in the study of motion and physics. Even though we can't show a perfect free fall on Earth due to the effects of air resistance, we can approximate free fall motion. One of Galileo's key findings is that all objects accelerate equally regardless of weight. Objects in free fall get faster as they fall in a linear fashion. On a distance-time graph, you get a parabola, implying the object gains more distance in successive time intervals.

Earlier in this chapter, I discussed the force created by gravity. This concept is called *weight*, and it will be the topic of the next chapter.

CHAPTER 4

Getting Heavy: Weight

IN THIS CHAPTER, YOU'LL LEARN MORE ABOUT the effects of gravity. Take an object in your hand. What do you feel? Do you feel the object pushing down on your hand? This push is called the object's WEIGHT. Weight is the FORCE of gravity pulling downward on the object.

This force is partially based on the MASS present in the object. Mass is a fundamental property of MATTER; it is a measure of the physical matter in an object. The basic measure of mass in the metric system is the gram or, more commonly, the KILOGRAM (which equals 1,000 grams). A 2-kilogram weight in a set made of the same material is twice as massive and large as a 1-kilogram weight because it contains twice as much matter. A larger object does not, however, necessarily have more mass than a smaller object. An inflated balloon, for instance, may be larger (that is, it has more volume) than a steel weight, but the balloon is not necessarily more massive.

So *weight* is a measure of the effect that gravity has on mass. Weight is a force directed downward toward the center of our planet. It is always considered an attractive force between the mass of the object and the mass of the planet. The Earth's mass pulls down on the object's mass, creating weight. But the Earth does not have to touch the mass physically to exert this force.

INTERESTING FACT

In contrast to weight, which is based on an attraction of a mass with the Earth, ELECTROMAGNETISM, the force seen in electricity and magnets, can either *pull* (*attract*) or *push* (*repel*). Both weight and electromagnetism are examples of FIELD FORCES: the object that generates the force does not have to come in physical contact with other objects to exert that force. Arrows called VECTORS are often used to show the direction that forces act.

Measuring Weight

Forces are examples of vectors that can be shown with arrows. The arrows point in the direction of the force, as shown here.

surface friction

air resistance

Wait, There Are Other Common Forces

Weight is just one common force. There are other types of forces, such as CONTACT FORCES, which act when two or more objects or surfaces touch. SURFACE FRICTION is a type of a contact force. The cars you built in Chapters 1 and 2 were affected by surface friction. AIR RESISTANCE is a frictional force too. Consider a race car: The surface of the road interacts with the surface of the tires to create *friction*. Air molecules make contact with the car to create *air resistance*. Friction can actually help us move! Walking on ice is hard because we don't have enough friction!

Then there are support forces, called NORMAL FORCES, from a surface like a table, floor, or desk, that often "hold up" the weight of objects. *Normal* is a math term meaning right-angle or PERPENDICULAR. The direction of the normal force is perpendicular to the surface of the table. For an object at rest on the desk, the normal force is directed upward and numerically counters the downward force of the weight of the object.

TENSION is a force found in strings, cables, and chains when these are pulled tight. You can cut a string, attach it to a weight, and then lift the weight by the string. As you hold the weight stationary, the string is pulled tight and acts to support the weight. The tension in the string is equal to the object's weight and that tension is directed upward while the weight is downward. Tensions and normal forces are important in structural design and architecture. Take the Golden Gate bridge, for example. The tensions in the cables and the normal forces from the piers

act together to support the weight of the bridge and the cars that drive on the bridge.

Rubber bands, springs, bungee cords, and other stretchy materials can create ELASTIC FORCES that are often used as support forces, too. Although similar to the tension you see in pulled string, the attachments between the molecules and atoms in a more elastic material can stretch a longer distance before breaking. Cars have springs that support the car's weight, as shown here. I can also support a mass with a rubber band.

Force can be quantified, too. In the United States, you use POUNDS (abbreviated lbs.) as the unit of measure for the force of weight. Most other countries in the world use the NEWTON. The Newton is the official metric unit of force and the most common unit of force used in physics. The Newton (abbreviated N) was named for Isaac Newton (1642–1727), the English scientist who researched forces and published the three famous laws of motion (more on this in Chapter 6). Since both the Newton and the pound are units to measure the amount of a force, you can convert one to the other.

Instruments to Measure Weight

Instruments are available to measure the force of weight. You can record this quantity in Newtons or in pounds depending on how the instrument is CALIBRATED. Some weighing instruments are common, like the bathroom scale, which measures the force (or weight) of gravity. Other tools are more specialized, like what you find in a physics classroom.

For instance, spring scales have a spring inside them that stretches due to an elastic force, with a length that is proportional to the force applied.

You can see some examples in Figure 4-1. What's meant by *proportional*? A string scale may be calibrated to stretch 1 inch if 10 lbs. of force is applied; it would then stretch 2 inches with a force of 20 lbs. applied. Spring scales have hooks for applying different weights or pulling with different forces. They are often colored and the colors indicate how much weight can be applied for that specific spring scale.

Another type of scale is the pan scale. You've probably seen them in grocery stores, so you can weigh out fruits and vegetables. They are also used

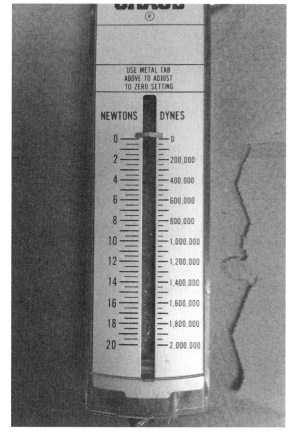

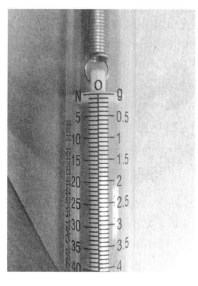

FIGURE 4-1 Various spring scales

FIGURE 4-2 Two types of pan scales

in the post office to weigh envelopes and packages for the proper postage. It is also possible to buy digital scales for weighing portions of food and other things. You can see some examples of pan scales in Figure 4-2.

An Equation for Weight

When you calculate an object's weight, you consider its mass (mass in kilograms) and its acceleration due to gravity (lowercase g in meters per second per second). When you multiply these numbers, you multiply the units, so $(kg \times m)/\sec^2$. These units become the Newton, which is much easier to write and remember. Here's the equation for weight:

$$W = mg$$

with weight measured in Newtons, mass in kilograms, and g, the acceleration due to gravity, measured in meters per second per second.

As mentioned, the Newton is the metric unit for weight and force. To convert to pounds, you can use 4.45 Newtons = 1 pound. Another conversion

is 2.2 pounds of weight (or force) = 1 kilogram of mass. If you take my weight in pounds (166.5 lbs.) and multiply by 4.45, my 166.5 pounds becomes equal to 741 Newtons. Dividing by 9.8 m/s/s or m/s², the acceleration due to gravity, I find that I have 75.6 kilograms of mass.

Try this out on yourself, if you're curious. Weigh yourself and then use the weight equation and conversion to determine your mass.

Project: Build the Trebuchet

Can you use weight to move an object? Have you heard of a trebuchet or seen the huge pumpkin-chunkin' trebuchets on TV? A *trebuchet* is a siege engine used in the Middle Ages to destroy castles.

 **NOTE**

There were many different types of siege engines. In the next chapter, you'll build a catapult.

A trebuchet works with a weight that drops downward, propelling the rock or object forward. The rock or object is situated at the end of a LEVER ARM in a sling. As the weight drops, the lever arm moves forward and upward, sending the rock flying out of the sling toward the castle. Medieval trebuchets were giant instruments of war with the ability to fling up to 350-pound boulders at castle walls! In the trebuchet shown at the top of Figure 4-3, large washers are used for the weight. As the washers drop down, the lever arm moves upward and forward. The projectile is placed in a sling that slides horizontally along the bottom railing and then moves upward and forward with the lever arm. Two pieces of string hold the sling. The other two ends of the string are looped and attached to the lever arm. One of the strings is

Famous Scientists

Robert Hooke (1635–1703) was a scientist engaged in research in many areas of science. The portrait you see here is an impression painted in 2007 by Rita Greer for Gresham College in London, where Hooke was a professor of geometry. The artist includes in this portrait a listing of Hooke's interests as well as a spring, symbolizing Hooke's research with springs and elasticity. Because of his many interests, he is called a *poly-math*, a person whose interests extend to many subject areas. In physics, Hooke investigated ELASTICITY. HOOKE'S LAW, which says that a substance stretches proportionally as weight is applied, is named for him. Hooke also studied gravity and the effects of gravity among the sun and planets. In addition to these discoveries, Hooke used the microscope, shown in the first photo, and discovered the *cell*, the basic building blocks of plants and animals. The second image is his drawing of the cell structure found in cork.

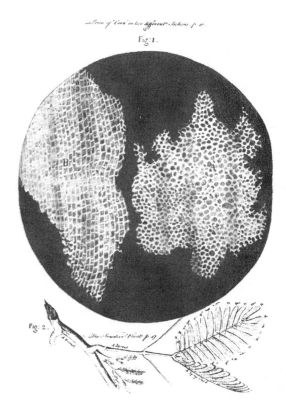

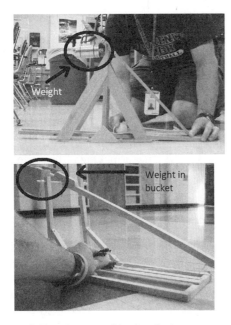

FIGURE 4-3 Two types of trebuchets

designed to release when the lever arm is at the top of its motion. The other string stays attached to the lever arm. The ball is thrown out of the sling and moves forward and upward. By changing the weight and the string attachments, you can launch different types of projectiles at various speeds and launch angles. The trebuchet shown at the bottom of Figure 4-3 has weight added to the hanging basket. You can add different types and amounts of weight to the basket. The firing mechanism is the same as with the top trebuchet.

Building a trebuchet is a fun project. You can find plans and kits on the Web or in some hobby stores. The two trebuchets in Figure 4-3 were made from kits. For this project, you'll build one from scratch that doesn't require buying much or using many tools. This trebuchet is ideal for flinging marbles several yards.

NOTE

I found the idea for this trebuchet at Instructables.com and modified some of the instructions. You can find more resources in the "Resources" appendix.

Things You'll Need

Here is a list of the things you'll need to build your trebuchet:

Parts

- **Five plastic rulers**   Make sure you get the ones with the holes in them. Four of these will serve as the base for your trebuchet and the other will be the lever arm.
- **³⁄₁₆" dowel rod**   Dowel rods are available at hobby stores in one-yard lengths. You won't need the full yard, but a width about two inches wider than your box.
- **Weight**   You'll link the weight to one of the holes of the ruler so you need a way to attach the weight to the ruler. I used a set of keys, as the keys had a key ring that made it easy to attach it to the ruler with a paper clip.
- **Six sink washers**   These will go on your dowel rod to keep it from shifting side-to-side. Mine were about ⁵⁄₈" in diameter with a ³⁄₁₆" hole in the middle (same size as the dowel rod) and made of rubber.
- 1 yard of string
- A shoebox or similar-sized box
- A strip of cloth approximately 3 inches long by 2 inches wide. I used an old shirt.

Tools

- Scissors
- Hot glue and hot glue gun
- Masking tape or duct tape
- Extra ruler for measuring
- Pencil
- Hand saw or hobby knife
- Small ball or marble
- Small nail
- Paper clip

BE CAREFUL!

Always make sure an adult is supervising when you're working on the projects. The tools needed for this project include a hobby knife and hot glue gun, both of which can cause injuries if used improperly. Remember, the completed trebuchet will throw a small object and should not be pointed at anyone.

Steps to Build the Trebuchet

1. The first step is to build one side of the base of the trebuchet. Line up the large hole at the top of two rulers with a pencil and angle the bottoms of the rulers against the length of the box.

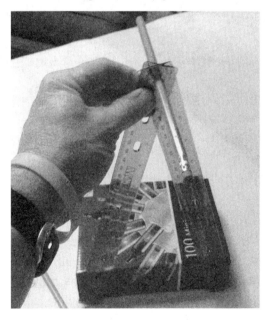

2. Apply tape to the bottom of the rulers to attach them to the box securely.

3. Now you're are ready to line up two more rulers to make the other side of the base. Use your pencil to line up two more rulers on the other side of the box with their top holes aligned with the other two rulers you just taped down in Step 1. Now your base is completed.

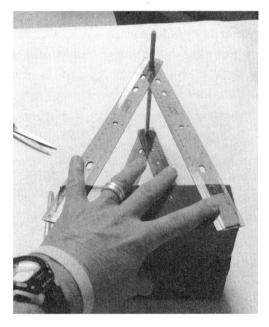

4. You're ready to cut the dowel rod. Set the dowel rod on the box, as shown next. Cut a

piece, leaving several extra inches of dowel rod on each side of the box.

 HINT!
....
I used a spare ruler to make my measurement and cut the dowel with a small hand saw.
....

5. Put the dowel rod through the large holes at the top of the four rulers attached to the base. (I placed the dowel rod in the second oval hole of the fifth ruler shown here.) This fifth ruler acts as the lever arm of the trebuchet and is located in the center of the box.

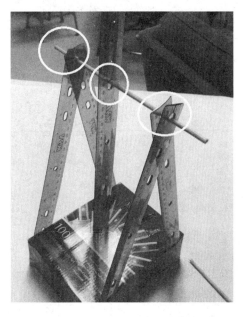

6. When you have everything lined up, remove the dowel and add the sink washers, and then reinstall the dowel with the sink washers through the correct ruler holes. Move the dowel and sink washers until the rulers are snug. Make sure the lever arm ruler can rotate around and is not too snug.

7. Bend a paper clip so it goes through the large circular hole toward the end of the lever arm closest to the dowel. Attach the weight (I used a ring with keys on it).

8. Now you're ready to construct the sling. This sling is constructed for a ball the size of a marble. It will hold the ball as you pull the lever arm around and then release the ball so it flies forward as the weight drops. For the sling material, I used a piece of an old T-shirt that was about 3" long and 2" wide. These dimensions don't have to be exact.

9. Cut two pieces of string, each one about 18" long.

10. Place the two strings on the cloth, leaving a small length of cloth to the left and right of the string. One string is at the left on the cloth with about 3" of the string above the top of the piece, as shown next. The other is on the right

side of the cloth with about 3" of the string below the bottom of the cloth. The two ovals in the picture represent the ends of the string.

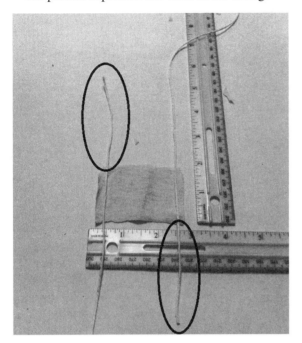

11. With a hot glue gun, apply a line of hot glue on the right string and then fold the small length of cloth over the string. This creates a sealed enclosure and secures the string on the cloth. After the right side has sealed, repeat this step on the left side.

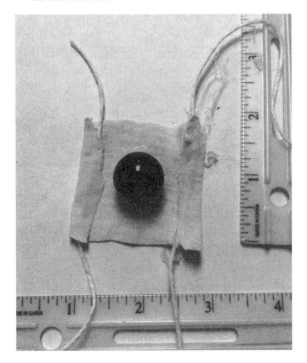

BE CAREFUL!

Hot glue can get *very* hot. Be careful when using it!

12. Starting with the left side, tie the end of the string above the cloth to the bottom part of the left string. Make sure your knot is tight. Repeat the same step with the right side. You're creating the pouch-like enclosure that will hold the ball firmly until the lever arm releases the ball. Here you can see the two knots.

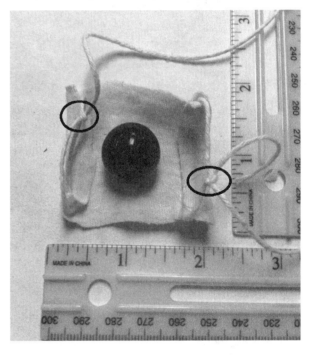

13. Attach a small nail with duct tape to the opposite end of the lever arm ruler with the weight on it. The nail should stick out the end

of the ruler about ¼ inch and have a slight upward angle, as shown here.

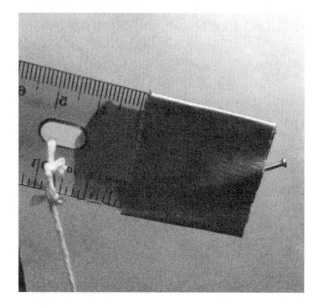

16. Take the left end of the string on the pouch (the string without the loop) and put it around the oval of the lever arm ruler next to the nail. Tie it in a knot around the oval so it's securely attached to the ruler.

14. With the pouch in front of you and the strings laid out to the left and to the right, cut the left side of the string 7" from the knot.

15. Cut the string on the right side of the pouch about 7" from the knot, but in this case, make a small loop on the end of the string.

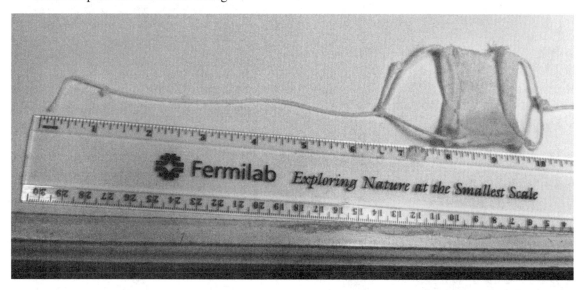

17. Take the end of the string with the loop and loop it around the nail.

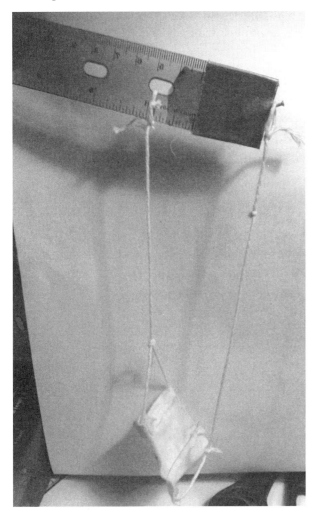

You're ready to test fire your trebuchet!

Fire Your Trebuchet!

1. Load the pouch with the small ball or marble, as shown here.

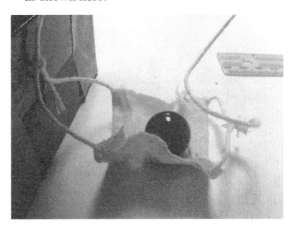

2. Drag the pouch with the marble back along the center of the box. As you drag the marble and pouch back, the side of the lever arm with the keys goes up and the side with the nail comes down.

3. When you release the pouch with your fingers, the ball and pouch should slide down the center of the box as the keys are pulled down. The lever arm comes up, making an arc. Your ball should release at the point where the lever arm ruler is almost vertical and the weight is almost at the bottom. The loop of string should come

off, which opens the pouch, releasing the ball so it flies upward and outward.

ruler_trebuchet_load_and_launch_closeup.avi

If your trebuchet does not firing correctly, there could be several problems. Here are a few tips:

- The weight you are using is not heavy enough. Make sure your ball is small and will fit in the pouch and is not too heavy.
- If the ball is released behind the trebuchet, it's being released too early.

- If the looped end of the string on the nail comes off before it should, you may need to adjust your nail or your string length. This happened to me a few times, and I adjusted the loop and pushed the nail so it had more of an angle.
- If your ball gets shot but lands in front of the trebuchet downward (without going up and out), it means the ball is being released later than it should. Again, you might need to adjust the nail and the string so the loop comes off at the correct time.

HINT!

The release of the ball can be a bit tricky. You may have to fiddle with it to get the ball to release at the right time.

See the website for a video of the trebuchet in action. Here is a sequence of screenshots from the video. The black oval shows the position of the marble.

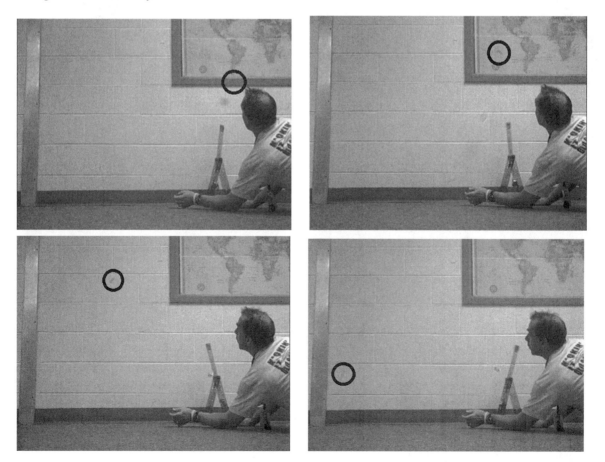

Summary

How did your trebuchet work? The trebuchet demonstrated that the force of weight can be used to send an object flying upward and outward. Forces are powerful things in our life. This chapter was all about basic types of forces: weight, tensions, and normal forces. You learned about mass and how to calculate the weight of an object. Normal forces and tensions are often used to support weight. The trebuchet sends the marble outward and upward. The flying marble is called a projectile, and the type of motion given to the marble is called projectile motion. You'll explore projectile motion in more detail in the next chapter.

CHAPTER 5

Storming the Castle: Projectiles

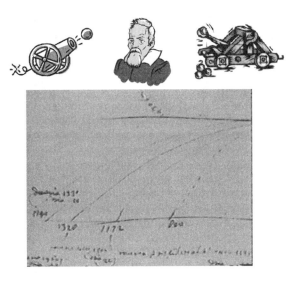

GALILEO INVESTIGATED PROJECTILE MOTION, too—the topic of this chapter. His experiments in projectile motion were a natural extension of his work with free fall and inclined planes. He let balls roll off inclined planes and fly through the air (Figure 5-1). By repeating the experiment many times, he realized that the path, or TRAJECTORY, followed by a ball through the air is a parabola.

Galileo was interested in the subject of projectiles because of his work in BALLISTICS. Ballistics is the study of projectiles and their trajectories, usually in the form of bullets, artillery shells, bombs, and cannon balls. As a consultant to Venetian military in northern Italy, he analyzed cannon trajectories and researched the speeds of the cannonballs, along with firing angles and how air resistance affected the motion of the projectiles. Figure 5-2 shows a page from Galileo's manuscript concerning trajectories.

Even before Galileo's time, many weapons of war were designed to shoot projectiles at castles. In the last chapter, you built a model of one such weapon, a trebuchet, and used a falling weight to fling a marble. In this chapter, you'll build a different kind of projectile—a catapult—and use video analysis to unravel the mystery of projectile motion.

What Is a Projectile?

A PROJECTILE is an object thrown or projected in some way. A thrown object usually has no ability to propel itself but must be given its motion by some external force. Once thrown, however, the object moves under the influence of gravity. A thrown baseball or football is a projectile. In fact, many sports involve putting objects into projectile

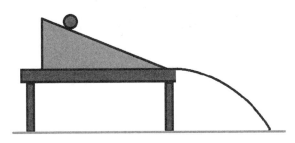

FIGURE 5-1 The ball rolls off the inclined plane and through the air, hitting the floor; this is projectile motion.

FIGURE 5-2 A page from Galileo's manuscript concerning trajectories

motion. You might kick a soccer ball toward a goal, shoot a basketball toward a hoop, or hit a tennis ball with a racquet; those balls in motion are projectiles. In sports like diving or gymnastics, divers and gymnasts put themselves into projectile motion when they jump off the diving board or dismount from the balance beam or uneven bars.

The path of a projectile represents up and down (*vertical*) motion as well as forward (*horizontal*) motion all wrapped into one trajectory. Both motions occur at the same time. Consequently, PROJECTILE MOTION is an example of TWO-DIMENSIONAL MOTION. And when you diagram the pathway or trajectory of a projectile, you'll find, as Galileo did, that it is a parabola!

Project: Build a Catapult

Catapults were devices invented in antiquity to project rocks, dead animals, arrows, and so on into the air. Often called SIEGE ENGINES when attacking a castle, catapults were also called *ballistas, springalds, mangonels, onagers,* and *trebuchets,* depending on how they worked and the projectiles fired. Most of them involved pulling a cord or rope tightly and allowing the tension to propel the projectile forward and upward. This is the type of catapult you'll build in this chapter.

Many different catapult kits are available in hobby and department stores, as well as online. Here, I show you how to build a catapult from scratch using easy-to-find parts and tools, but you can build one from a kit if you prefer.

Things You'll Need

Parts

- **Basswood**   This material is light and strong, and you can find it in hobby shops. I used an 8" long, 3" wide, and ⅜" thick precut piece of basswood for the catapult's base. You could use a similarly shaped piece of foam board or cardboard instead.

- **Square wooden dowel for the lever arm**   It should measure ½" × ½" and be long enough so you can cut a 12" length.

- **Six more square dowels for the support arms and bracing**   These pieces should be ½" × ½" and between 7" and 8" long.

- **Round wooden dowel**   The dowel should be ¼ or ⅜" wide and 6" to 7" long. You'll use this dowel as the rotational axis for the lever arm.

- **Several rubber bands**   I recommend #33 rubber bands.

- **Two eyehooks**   You don't need an exact size.

- Small plastic spoon

- Ping-Pong ball

Tools

- Eye protection (when drilling and shooting the catapult)

- Ruler for making measurements

- Pen or pencil for marking on the wood

- Drill and drill bits

- Hot glue gun and hot glue sticks

- Hobby saw and mitre box like the one shown here

 HINT!

You can also use a hobby knife, but the miter box and saw makes it easier to create 45-degree cuts for the catapult's support arms.

BE CAREFUL!

The projectiles launched by catapults can injure your eyes. Wear safety glasses when building them and shooting them. The tools used to build catapults—whether from scratch or a kit—involve using tools like saws, hobby knives, and hot glue guns. Be careful when using these tools. Make sure an adult is present to supervise and help, if needed.

Steps to Build Your Catapult

1. Choose a rectangular slab of basswood for the base of your catapult. You could substitute cardboard or foam board for the basswood, although it won't be as sturdy.

2. Take one of the rectangular dowel pieces (½" × ½" × 7 or 8") from the basswood package and cut off a 3" piece using the saw and mitre box.

3. Glue the 3" piece onto the base at one end using hot glue. Match it up with the width of the rectangular slab.

4. Now you'll create the support beams for the catapult. You need two rectangular wood dowels. These dowels should be ½" × ½" × 8" long.

5. Mark three holes in these two support beams. Line up the two beams and make three marks with a pencil or pen 1" apart along one side of the beam, starting 1" from the end of the beams.

6. Drill a hole through all six marks. The hole should be big enough for your circular dowel rod to slide through and large enough to allow for easy rotation. (I used a ⁷⁄₃₂" drill bit to create my six holes.) Drill on a block of wood so as to not damage your working surface, and drill straight and carefully and be sure to use eye protection.

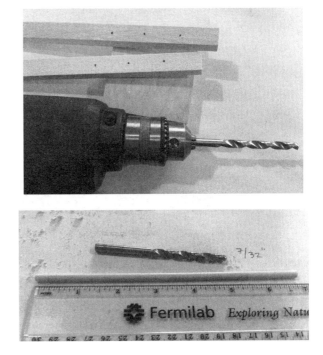

7. Mount the dowels onto your base. Find the middle of the base with a ruler and then hot glue the two support beams vertically along

each side of the base. Make sure the two beams are straight up and down. The three holes should match up; work your circular dowel through each to make sure the holes line up correctly.

8. Take two more of the ½" × ½" × 8" long square wooden dowels. You'll cut these into four angled pieces, each about 4" long. These dowels will serve as the support structures for the two beams.

9. Set up the mitre saw box to cut each dowel in half at an inward 45-degree angle (A). Cut the dowel (B), and then rotate the wood rectangular piece and make another angular cut (C).

A

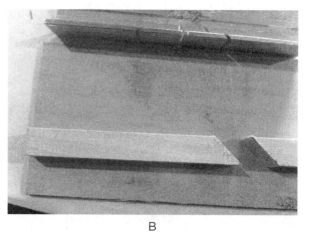

B

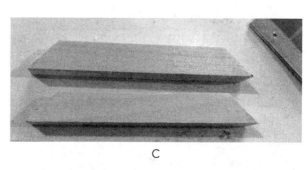

C

10. Glue these four angled supports onto your beams using hot glue.

BE CAREFUL!

Hot glue guns are dangerous! Be careful not to burn yourself!

11. Take a 5" piece of the ½" × ½" wooden rectangular dowel and straight cut a 3" piece, so you have a 2" piece leftover. You'll use both of these pieces.

12. Take the 2" piece and hot glue it to the inside of the support beams right above where the angular beams are attached. Hold this tightly while it dries. Then glue the 3" beam to the top of the two support beams, as shown here.

13. To construct your lever (swinging arm), take the long rectangular ½" × ½" wooden dowel that you purchased specifically for the lever arm and cut it to a 12" length using a straight cut.

14. Drill holes into the lever arm with the same drill bit as you used in step 6. (I used a $\frac{7}{32}$" drill bit that allowed my circular dowel to fit easily through the holes. I constructed the holes at the 4", 5", 6", 7", and 8" marks from one of the ends, as shown here.)

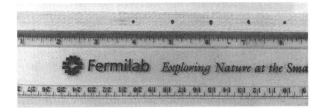

15. On one end of the lever arm, use a rubber band to securely attach the plastic spoon. On the other end, screw in one of the eyehooks.

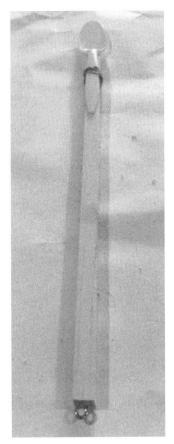

16. Attach the other eyehook to the catapult base, as shown here.

17. Now you're ready to attach the lever arm to the catapult base and support beams. Slide the circular dowel through the two support beams and through the middle hole of the lever arm. The lever arm will rest between the two beams and should rotate freely with little friction (Figure 5-3).

18. Cut a rubber band into one long piece. Then run the rubber band through the two eyehooks. Knot each end securely around the eyehooks (A). The rubber band should stretch as you pull

the spoon-side of the lever arm down (B). This stretched rubber band will give your Ping-Pong the catapulting speed.

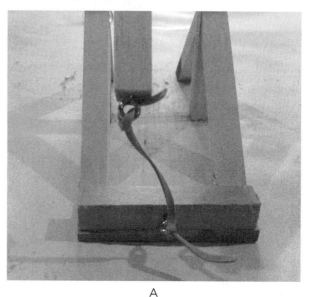

A

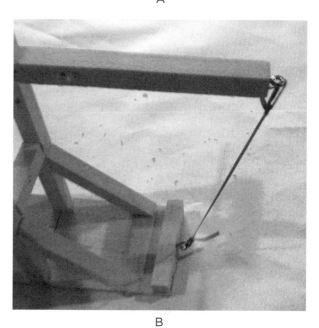

B

19. Try out the catapult yourself. Put the Ping-Pong ball on the spoon, as shown next, and stretch the rubber band by pulling down on the spoon-side of the lever arm. Hold the base, release the lever arm, and the Ping-Pong

FIGURE 5-3 Slide the circular dowel through the two support beams and lever arms.

ball should go flying! You may need to make minor adjustments with your spoon or other components.

 HINT!

Place the circular dowel through the different holes in the support beams and the lever arm to set up different release points and trajectories. My catapult launched the Ping-Pong ball several meters!

Experiment
Doing Physics with Your Catapult

Now it's time to explore the physics of our catapult launch. Remember, a projectile moves horizontally and vertically at the same time. But how will the Ping-Pong ball move horizontally and vertically? Will it move at a constant speed? What is its rate of acceleration? To analyze the motion of the Ping-Pong ball launched by the catapult, you'll film the motion of the ball as it moves through its trajectory after it is launched by the catapult. Then you'll use Tracker Video Analysis to figure out what's happening. You can graph both the horizontal and vertical motion. The types of graphs you create will reflect the type of motion the projectile has as it flies through the air.

If you don't want to make a video of your own, view the one I created by downloading it from the book's website.

Film Your Video

I built my catapult to launch a Ping-Pong ball. Make sure you launch a ball you can see on your video. A disadvantage to shooting Ping-Pong balls is the air resistance and wind. With too much air resistance, your graphs and results won't be correct. To minimize the effects of the wind, I videotaped my catapult in the garage. Make sure you move your video camera to a place where you can see the whole trajectory after the catapult launches the ball. Use a meter stick or defined length in your field of view. If your catapult is like mine, it can launch the ball pretty far. To get a full launch on video, don't pull the spoon and ball back so far.

Analyze Your Video with Tracker

Once you've created and downloaded your video to your computer, you're ready to analyze it using the Tracker Video Analysis software. Load your video in Tracker program and find the correct frames to match the throw from your catapult. Using my video as an example, I used frame 81 as my starting frame and frame 105 as my ending frame. I used a step size of 2, which gave me 13 different data points.

Select a calibration (meter) stick. I used a 100-cm meter stick in my video. Choose **Point Mass** and frame out your video to create your data points. Set up the coordinate axes to match you first data point (0). Make sure your first x and y points are both 0. Figure 5-4 shows a screenshot of my video in Tracker.

Set up Tracker to show the following columns: time (t), forward distance (x), up and down distance (y), forward speed (v_x), and vertical speed (v_y). Once

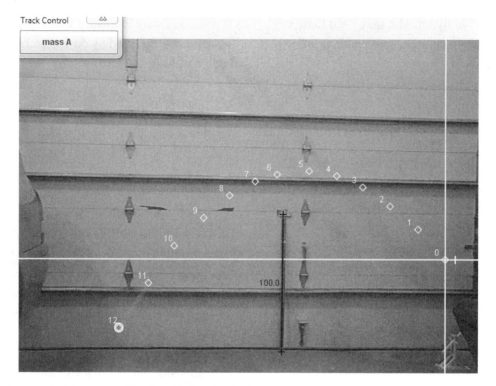

FIGURE 5-4 Screenshot from my Tracker Video Analysis

you have all five columns of data in Tracker, cut and paste each column into Excel, as shown in Figure 5-5.

Creating Graphs in Excel

The first graph you'll create is a horizontal (*x*) distance-time graph. Set up Excel to show the time on the *x* axis and the horizontal distance on the *y* axis. This represents the forward motion of

the Ping-Pong ball as it moves along its trajectory. Depending on the version of Excel you're using, the steps to create a graph might be slightly different. Basically, you select different columns depending on the graph want. For this graph, select the **t time (seconds)** and **x distance (cm)** columns and then create a scatter graph. A scatter graph is used to show trends in data.

Catapult built from scratch				
Tracker data				
time (seconds)	x distance (cm)	y distance (cm)	x speed (cm/sec)	y speed (cm/sec)
0	0	0		
0.067	-20.407	21.938	-309.623	290.511
0.133	-41.324	38.773	-309.623	229.351
0.2	-61.731	52.548	-298.156	164.368
0.267	-81.118	60.711	-301.978	87.918
0.334	-102.036	64.282	-336.381	7.645
0.4	-126.014	61.731	-305.801	-57.338
0.467	-142.85	56.63	-271.398	-114.675
0.534	-162.236	46.426	-294.333	-198.77
0.601	-182.133	30.1	-313.446	-275.221
0.667	-204.071	9.693	-313.446	-355.493
0.734	-223.968	-17.346	-317.268	-447.234
0.801	-246.416	-49.997		

FIGURE 5-5 My Tracker data pasted into Excel

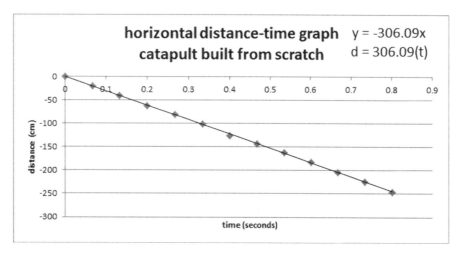

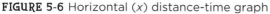

FIGURE 5-6 Horizontal (*x*) distance-time graph

Notice that the horizontal distance-time graph in Figure 5-6 shows a nice diagonal with a negative slope of 306.09. Because distance-time is a diagonal line, this means the Ping-Pong ball is moving forward at a constant speed. The number is negative because I filmed my catapult shooting the Ping-Pong ball from right to left. The speed 306.09 is in centimeters per second, or 3.0609 meters per second. This speed represents the constant speed forward for the Ping-Pong ball's trajectory. You can change the equation to match the physics variables:

distance = slope*[*time*] or *d* = 306.09(t)

The second graph you'll make is a vertical (*y*) distance-time graph. Set up Excel to show time on the *x* axis and the vertical (*y*) distance on the *y* axis. This graph should look much different than the one in Figure 5-6. The parabola in Figure 5-7 shows that the ball is moving vertically at a constant rate of acceleration. The Ping-Pong ball is also moving vertically as an object in free fall.

The third graph you'll create is a vertical (v_y) speed-time graph. Set up Excel to show time on the *x* axis and the vertical speed (v_y) on the *y* axis. The diagonal line on the graph in Figure 5-8 suggests,

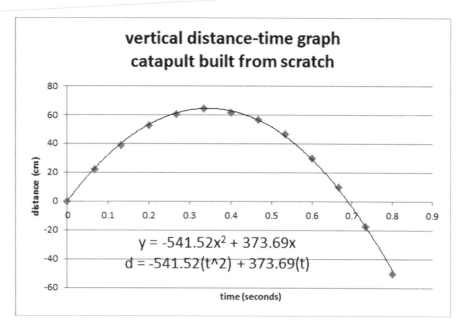

FIGURE 5-7 Vertical (*y*) distance-time graph

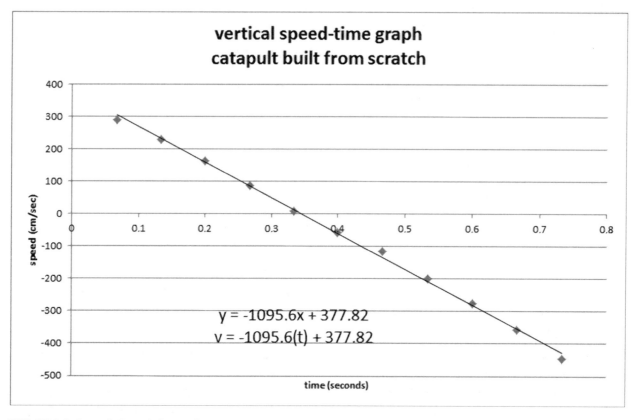

FIGURE 5-8 Speed-time (v_y) graph

as did the parabola in 5-7, that the Ping-Pong ball moves vertically in free fall. The slope is a measure of this acceleration and should be equal or close to –980 centimeters per second per second or –9.8 meters per second per second. Why does this occur? A projectile moves vertically under the full influence of gravity and –9.8 is the acceleration due to gravity on Earth! My number is off a bit (–1095.6 centimeters per second per second) but it's still close.

You've done it! By using your Tracker Video Analysis software and Excel you've analyzed your catapult video and determined how the Ping-Pong ball is moving. The Ping-Pong ball is a projectile and has both motion that is forward or horizontal in nature and vertical (up and down). Both of these motions occur simultaneously to form a trajectory or pathway. Using Tracker, you can break the trajectory apart to look at the two motions separately:

- Horizontally, the ball moves forward at a constant speed so you see a diagonal on the horizontal distance-time graph.

- Vertically the ball is in free fall, experiencing the full effects of Earth's gravity, so you see a parabola on the distance-time graph and a diagonal on the vertical speed-time graph. The slope of the diagonal is approximately little g, or around –9.8 meters per second per second.

If you enjoyed this experiment, try the same experiment with the trebuchet you built in Chapter 4.

Summary

Knowing how projectiles move is an important concept in understanding the physics of motion. It ties together two types of motion into one trajectory. If you can minimize the effects of air resistance, then the projectile moves horizontally at one constant speed. While it moves forward, it also moves vertically. Vertically, it is in free fall. Free fall is a constant acceleration that represents the effect of gravity. On our planet, this is *little g*, which equals –9.8 meters per second per second.

The catapult is a great device for flinging a ball into projectile motion. I hope you had fun building the catapult—and found it just as amazing to understand the physics of what is happening when the catapult releases the ball!

Famous Scientists

Guidobaldo del Monte was an Italian mathematician, physicist, and astronomer born in 1545, about 19 years before Galileo. He is important in the discussion of physics because he helped Galileo and supported his scientific endeavors.

When Galileo was a young man, he needed to find a job to support himself and his family. He sent Guidobaldo a scientific article he had written. Guidobaldo found the article Galileo sent him interesting and promising, so he contacted his brother who had connections with the University of Pisa in Pisa, Italy. Galileo was hired as a professor of mathematics in 1589. Later Guidobaldo also helped Galileo in securing a position at a different university, the University of Padua in 1589.

Guidobaldo was an important scientist and inventor, too. He developed drafting instruments to measure angles. Galileo incorporated one of his drafting instruments into a new tool called the *geometric and military compass.* Galileo developed this device for surveyors and cannon operators so they could calculate angles for certain trajectories. For cannon gunners, the compass was a much better way of changing the angle and elevating the cannon barrel. The gunner would put the compass in front of the cannon's barrel and move the side arms of the compass to measure a certain angle. The barrel could then be moved to match this angle.

This is a photo of Galileo's military compass. This compass is believed to have been constructed around 1604. The two side arms could pivot, offering different angular measurements.

CHAPTER 6

Acceleration: Newton's Laws of Motion

SO FAR WE'VE TALKED ABOUT SEVERAL different types of motion using ideas developed by Galileo. To better understand *why* objects move as they do, however, you have to look at the discoveries of another great scientist: Isaac Newton, an Englishman, who was born in 1642 and became a professor at Cambridge University in 1669. Like Galileo, Newton made a wide range of discoveries that we still use today in mathematics and physics.

certain conditions and predicts what *should* happen if these conditions are met. In the early 1600s, Johannes Kepler published his three laws of how planets move around the Sun: the law of orbits, law of areas, and law of periods. These laws describe and predict the motion of the planets as they journey in their orbits around the Sun.

Newton organized his explanations of motion into what are now called Newton's three laws of motion.

Understanding Why: Newton's Laws of Motion

Scientists have hypotheses, theories, and laws. As you learned in Chapter 1, a hypothesis is a statement or question that can be tested in an experiment. For example, you might hypothesize that an object in free fall accelerates constantly as it falls. You can test this hypothesis by conducting experiments.

A THEORY summarizes one or more hypotheses. A theory has been rigorously tested many times and is considered valid until it is disproved. You may have heard of the theory of relativity. Proposed by Albert Einstein in the early 20th century, it attempts to explain, among other things, how gravity works.

A scientific LAW generalizes a set of observations. A law of motion describes how objects move under

Newton's First Law

NEWTON'S FIRST LAW says the speed (velocity) of an object will stay constant unless an external force acts on the object. If the object is not moving, it will remain motionless unless a force acts on it. If the object is moving at a constant speed, then a force is needed to change this speed. Newton's first law is often stated another way: a body in motion tends to stay in motion; a body at rest tends to stay at rest. A force is needed to change what it is doing. The catapult you built in Chapter 5, shown here, works because of Newton's first law.

The lever arm of the catapult and the not-so-firmly-attached Ping-Pong ball move as one unit. The wooden rod quickly stops the lever arm, but the Ping-Pong ball continues on in the same motion because it was not firmly attached to its holder.

The importance of wearing seat belts can be nicely illustrated using Newton's first law. If you're in a car going 70 miles per hour on the freeway and hit something, the car will be stopped suddenly like the arm of the catapult. But, like the Ping-Pong ball, you will continue the car's motion at 70 miles per hour—unless a seat belt stops you along with the car. The seat belt, an example of Newton's first law, keeps you "attached" to the car.

Applying the same law, an object at rest can be difficult to move because it resists a change in motion. This is called INERTIA and Newton's first law of motion is also called the LAW OF INERTIA. It applies equally to objects in motion and at rest. Objects with lots of mass have lots of inertia, which is why it's easier to push a wagon than a car!

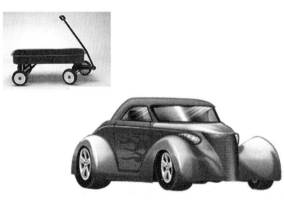

Galileo's writings inspired Newton to compose his first law. Remember the inclined plane discussed in the last chapter? Galileo used this to study projectile motion as well the motion of the ball along the ground after it came off the ramp. Galileo reasoned that a ball rolling off of one of his inclined planes would maintain its motion along the ground if not for friction.

Newton's Second Law

The ball rolling off the ramp is a good way to introduce us to NEWTON'S SECOND LAW. Newton's second law says an unbalanced, or net, force causes an object to accelerate. You can write it as a mathematical equation:

$$F_{net} = ma$$

You can read the equation as "*force* equals *mass* times *acceleration*, or *force = mass × acceleration*." An UNBALANCED FORCE or NET FORCE causes the mass of an object to accelerate. Figure 6-1 shows a ball rolling off of Galileo's inclined plane with an unbalanced (net) force, which is the friction on the rolling ball.

SURFACE FRICTION is a force that can cause a moving object to slow down and eventually stop. Once the ball comes off the ramp, no other force is propelling the ball forward. Friction is an unbalanced or net force. This unbalanced or net force creates an acceleration via Newton's second law ($F_{net} = ma$). The ball has mass and that mass accelerates (negatively) because of friction. In this case, the acceleration opposes the ball's speed and its speed gets slower until finally the frictional force stops the ball.

What about the fan car you built in Chapter 2? The fan car accelerates when you turn it on. Its speed gets faster and faster from a resting state. Consequently, there must be a net or unbalanced force.

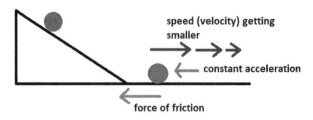

FIGURE 6-1 The unbalanced force of friction causes the ball to accelerate. In this case, the acceleration opposes the velocity (ball's speed) as friction slows the ball down.

In Figure 6-2, in the image on the left, when the fan car accelerates, the car goes faster and faster, which means a net, or unbalanced, force is acting on the mass of the car. The fan engine propels the car forward. This is the force from the engine. The atmosphere pushes back as the car moves through it. This force is called AIR RESISTANCE, or DRAG. Friction between the wheels and the surface creates a frictional force. In this case, the force of the engine is greater than the sum of the drag and frictional forces. This creates a net force, and this net force creates the acceleration. Figure 6-2 is called a FORCE DIAGRAM or FREE BODY DIAGRAM because it includes the forces acting on the object.

But the fan car will eventually stop accelerating and move at a constant speed. This means that the acceleration equals 0 and that the net force must also equal 0 due to Newton's second law. The frictional force doesn't change numerically, and the engine supplies the same force. As the fan car's speed increases, however, the effects of the atmosphere become greater, increasing the drag.

In the photo on the right in Figure 6-2, the force from the atmosphere (drag) has a larger arrow because the car's speed has increased. Objects moving faster have a greater impact on the atmosphere against them than those moving at a slower speed. Now the forward force from the engine is balanced with the sum of the frictional and drag forces. Because the forces are balanced, the net force = 0. If F_{net} = 0, then the fan car's acceleration is zero and, consequently, the car won't accelerate more. It maintains a constant speed.

Newton's Third Law

A force is a push or a pull that is an interaction between two objects. NEWTON'S THIRD LAW says for every action there is an equal and opposite reaction. If I push on the wall, the wall pushes back. The hand pushing the wall is the *action*. But flip the nouns around: the wall pushes the hand; this is the *reaction*.

You can see this law in action with a set of spring scales, like those shown here.

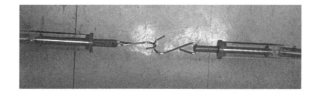

My left hand pulls against the spring scale in my right hand, and my right hand pulls against the

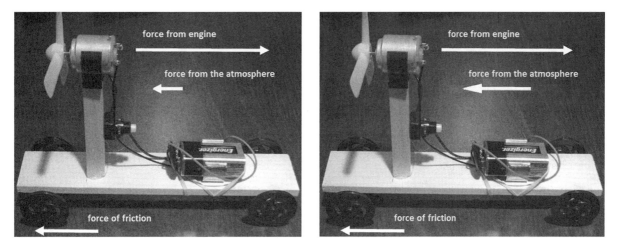

FIGURE 6-2 When the fan car accelerates, you have unbalanced forces, as shown in the image on the left; when it moves at a constant speed, the forces are balanced, as shown in the image on the right.

spring scale in my left hand. Both scales have the same value. Left hand pulls the right hand; right hand pulls back against the left hand. One of the forces is directed toward the left and the other force is directed toward the right.

You can graph this by using two dual-range force probes that can be attached to the computer to create a force graph. I hooked two of them together, just as I did with the spring scales.

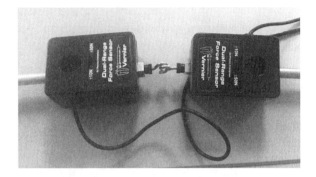

As with the spring scales, I pull with my left hand and my right hand responds by pulling back. The force is equal in amount but opposite in direction. Here is a sample graph.

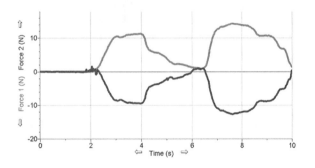

One of the spring scales registers its force as a positive force, shown as the upper line in the illustration, the other as a negative force, shown as the lower line. The positivity and negativity account for the left and right directions of pull.

Both forces are numerically equal, which makes the two graphs symmetric.

Newton's third law can be counter to what your intuition tells you sometimes. Take a look at this picture:

The car and the bug are moving toward each other. The bug hits the windshield of the car. And the windshield hits the bug. Both the bug and the windshield exert and experience *equal* forces on each other. Why is this confusing? The *effects* of the forces can be vastly different, and to investigate the effects of the force you need to use Newton's second law.

$$F_{\text{bug on windshield}} = F_{\text{windshield on bug}}$$

Place this equation for Newton's second law on each side of the equals sign:

$$M_{\text{windshield}} * a_{\text{windshield}} = m_{\text{bug}} * A_{\text{bug}}$$

The mass (*M*) of the windshield/car is much greater than the mass of the bug. Consequently, the acceleration (*A*) the bug feels is proportionally greater than the acceleration (*a*) on the windshield/car. The bug's body cannot survive this acceleration, and you can usually see the evidence on the windshield. But remember: the car *does* experience a very tiny acceleration (*a*) from the bug, and its speed is decreased by a very tiny amount.

Famous Scientists

As a young man, **Isaac Newton** (1642–1727) tried his hand as a farmer on the family farm in England. But Newton found farming monotonous. As a boy he enjoyed building things, doing chemistry, and studying astronomy. His mother finally relented and allowed young Isaac to go to college. Newton had very little money, however, and was only able to attend Cambridge University by waiting tables and cleaning the rooms of the teachers and wealthier students.

By 1664, he had done well enough at college to receive a scholarship for additional years of study. Unfortunately, the plague was spreading throughout Europe and officials decided to close the university. Students were sent home and Newton went back to his family's farm. For the next several years, Newton, now 23 years old, would turn his attention to motion and how to explain motion.

During this time, Newton discovered his equation for gravity, his laws of motion, and he developed mathematical processes, such as calculus, as tools to explain his ideas. A famous story says that Newton saw an apple fall to the ground on his farm and compared the falling apple with the orbit of the moon around the Earth. These observations led to Newton discovering how gravity works on objects here on Earth and in the solar system. Beyond these discoveries, Newton also invented a new type of telescope (a Newtonian reflector, shown here) and developed a theory of light and color, discovering that a prism breaks white light up into the color spectrum.

When the university reopened, Newton continued his studies. He became a professor and used his laws of motion to further Galileo's ideas. Galileo had come up with the physics of projectiles; Newton extended these ideas by adding the force of gravity. In the 1680s, Newton compiled his discoveries in a three-volume series of books called *Philosophiae Naturalis Principia Mathematica* (*Mathematical Principles of Natural Philosophy*). Isaac Newton was one of the greatest scientists and mathematicians who ever lived; many of his discoveries, including his three laws of motion, are widely used to explain many of the things we see around us even today.

Project: Build a Water Rocket

Water rockets work by pressurizing air, which then forces water out of the rocket's nozzle at a high rate of speed. This creates the force called THRUST that gives the rocket upward acceleration and speed. Building and launching a water rocket will help you understand Newton's three laws of motion. Newton's laws describe the forces that make rockets work—from the huge NASA rockets that send satellites into space to smaller model rockets. Rockets involve unbalanced forces, action, reaction, mass, acceleration, gravity—just about every concept we've talked about so far!

For this project, you need a water rocket and a water rocket launcher. You can buy water rocket launchers online for about the same amount of money that it costs to build one. Building a launcher from scratch is a bit tricky, and I'm going to recommend that you buy one instead. You can see the launcher I got in Figure 6-3. It *is* possible to build a launcher from scratch, though, and I list some websites that provide instructions in the "Resources" appendix if you want to give it a try.

Typically, you need to weight the launcher down with a rock, brick, or some other heavy object, or else you'll need to drive stakes into the ground to hold it. You also need a hand pump, bicycle pump,

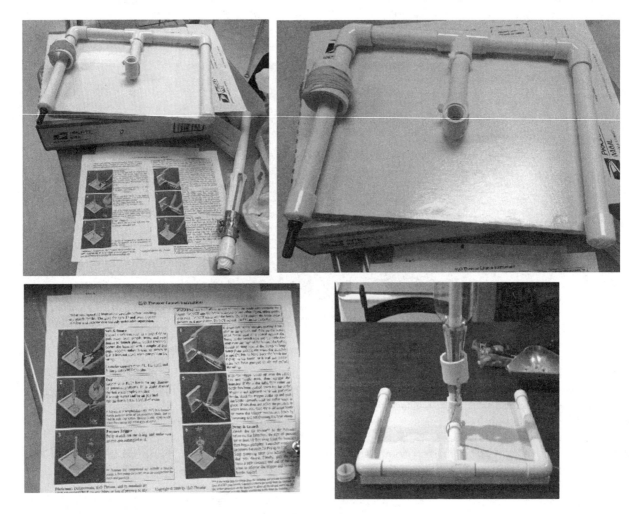

FIGURE 6-3 The water bottle launcher and supplied directions

or air compressor as an air source, preferably with an air-pressure gauge built in.

Here you will build the rocket that you will send aloft from the launcher. The rocket can be made from simple, easy-to-find materials and is a lot of fun to make. You can also customize it to look the way you want.

Things You'll Need

Parts

- **Two empty water or carbonated beverage plastic bottles**   0.5-, 1-, 1.25-, and 2-liter bottles work well. You need one for your rocket and one that's the same size for the nose cone. The bottles that seem to work the best are PET bottles. (PET is a type of plastic used in certain types of water and beverage bottles.)
- **Cardboard box, foam board, or thick paper**   You'll cut this to make the rocket's fins.
- **Ballast**   Something to add weight to the rocket like a ball of clay, an apple, orange, or even cat litter.
- Roll of clear strapping or shipping tape
- Quick drying epoxy glue (optional)

Tools

- Scissors or hobby knife
- Ruler

BE CAREFUL!

Take care with the scissors or hobby knife when you cut your fins and nose cone. Always make sure an adult is there to supervise when building and launching your rocket.

Steps to Build Your Water Rocket

I made two rockets of different sizes. You can choose a bottle size that suits you best. Let's get started!

1. Clean out the bottles and carefully remove the labels with scissors.

2. Cut off the retaining ring on each bottle with a pair of scissors.

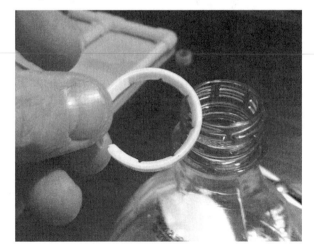

3. Use strapping tape or clear shipping tape to reinforce the bottle. Reinforcing the bottle is important as you'll be adding air pressure. I circled the bottles several times from top to bottom with shipping tape for extra reinforcement.

4. Cut the fins for your rocket. The rocket can have three or four fins. On my 1-liter bottle, I used four fins, and on my 2-liter bottle, I used three fins. I cut identical triangular shapes from a cardboard box. Other materials such as foam board or heavy paper could be used for fins, too.

5. Attach the fins to the bottle with additional shipping tape. You could also use epoxy to attach your fins. Make sure they are straight.

NOTE

The top of the bottle is the bottom of your rocket!

6. For the nose cone, take a bottle of the same size and cut it around six inches from the top of the bottle. Leave the screw lid on for the nose cone. The nose cone will fit onto the top of the rocket (the bottom of your bottle).

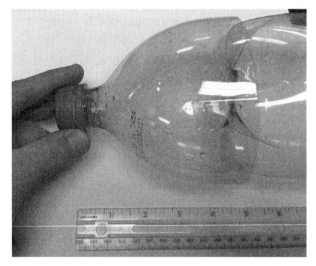

HINT!

Don't attach your nose cone yet as you need to add the weight first.

7. Put some weight inside the nose cone to keep the rocket stable as it flies. I decided to use an orange as it was about the right mass and would make the rocket easier to see and analyze on my launch videos.

8. Tape each cone to the top of the rocket securely with additional shipping tape. Finally, put the rocket on your launcher to make sure it fits before launch!

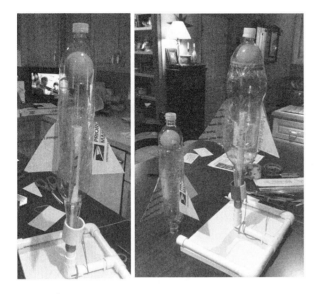

Liftoff! Launch Your Water Rocket

Ready to launch your rocket? You need water, your launcher, and an air pump at your launch site.

 HINT!

I recommend also bringing some basic tools (pliers, hammer, hobby knife, scissors) and some extra shipping or masking tape, as repairs are often needed. Bring a video camera if desired.

Find an open field or large parking lot. You need lots of space because water rockets can go very high!

 **BE CAREFUL!**

Launch your rocket well away from people, cars, windows, and so on. Make sure to read about water rocket safety here: http://www.sciencetoymaker.org/waterRocket/safetyWaterRocket.htm. Always have an adult present to give you a hand.

Fill your rocket about half full of water and place it on the launch pad. Lock it down with the launch pad trigger. (I had a bit of trouble with the launch pad trigger collar being loose and not remaining up over the cable ties of the launcher. I added additional strips of masking tape to the inside of the collar.)

Attach the air pump connector to the valve stem of the launcher. Use the gauge on your pump to get the bottle pressure to around 50 psi (of air pressure). Release the trigger following the instructions for your launcher to achieve rocket liftoff.

Figure 6-4 shows some cool pictures of my 1-liter rocket blasting off.

I had several good flights and took some video, which I then analyzed using the Tracker Video Analysis software. If you decide to video your launch and try the analysis, make sure you have the camera placed away from your launch pad and film as much of the launch as possible. I recommend a tripod for camera stability. But make

FIGURE 6-4 Blast off!

sure the camera is close enough so you can frame out the rocket moving upward (see Figure 6-5). I used the orange in each rocket as a spot to track on the video. The Tracker software can help you determine the rocket's acceleration and speed. The analysis can be hard as the water rockets are quick, but I obtained speeds of around 30 meters per second, which is getting close to 70 mph!

FIGURE 6-5 Framing out one of my water rocket launches in Tracker

the air in the water rocket, you force the water to come shooting out the bottle's nozzle. This creates a force that causes the rocket to move upward, illustrating Newton's third law: for each action, there is an equal and opposite reaction. This upward force is greater than the force of gravity or air resistance acting downward on the rocket. Consequently, an unbalanced force gives the rocket an acceleration ($F_{net} = ma$), illustrating Newton's second law: a net force acting on an object accelerates the object. This net or unbalanced force creates quite a display of acceleration, which is measurable with Tracker. This force changes the state of motion of the rocket from one of rest to one of increasing speed, illustrating Newton's first law: the speed of an object remains constant unless an external force acts on the object.

Building water rockets is a project you can take much further, if desired: you can find plans online for building multistage rockets, rockets with parachutes, drop-away boosters, and so on. Go to the book's website, where I've posted several videos of my blast-offs, along with my Tracker data and analysis. In the next chapter, I discuss several types of energy associated with the study of physics and how we can use these types of energy to explain some types of motion.

Summary

I hope you enjoyed this project! Water rockets make a great science project that incorporates all three of Newton's laws of motion. By pressurizing

Moving Forward: Kinetic Energy

ENERGY IS A WORD WITH MANY MEANINGS.
A lively person has a lot of energy. Energy is a power source for our houses and cars. And there are different types of energy, including solar, wind, gasoline, and hydroelectric. In this chapter, I discuss several types of energy associated with physics and how these types of energy can be transformed from one type to another. And to demonstrate these ideas, you'll build a mousetrap car, a car that "works" because of these energy transformations.

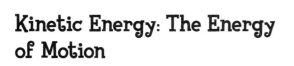

Kinetic Energy: The Energy of Motion

So far, you've studied the physics of several types of motion. There are different types of energy, too; and these types of energy can be related to moving objects. In this chapter, we'll look at three types of energy.

The first is KINETIC ENERGY. Kinetic energy is based on two things:

- The mass of the object
- How fast the object moves (that is, its speed, or velocity).

Kinetic energy is measured in a unit called the *Joule (J),* named after James Joule, a 19th century British scientist. The kinetic energy (abbreviated "KE") of an object is found with the following equation:

$$KE = \tfrac{1}{2}(m)(v^2)$$

which you read as kinetic energy equals ½ the *mass* of an object multiplied by its *velocity* squared.

If an object holds steady at one speed, then its kinetic energy stays constant. In Chapter 1, your constant-speed vehicle moved at approximately 1.1 meters per second (m/s). Knowing this, what would be the car's kinetic energy?

KE equals ½ the car's *mass* times its *velocity* squared. If the car has a mass of 0.4 kg and is moving at 1.1 m/s, KE = ½ (0.4 kg)(1.1 m/s)² = 0.242 Joules (J).

As long as your vehicle maintains its constant speed, its kinetic energy remains constant. What happens to the kinetic energy if the car's speed changes? The fan car you built in Chapter 2 (see Figure 7-1) started from rest (initial speed = 0 m/s) and eventually obtained a speed of around 0.8 m/s.

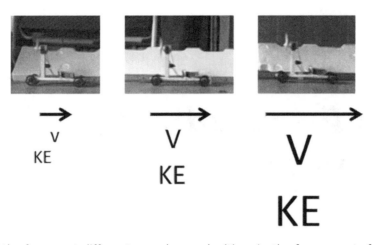

FIGURE 7-1 Images of the fan car at different speeds or velocities. As the fan car gets faster, its speed, or velocity (V), increases, which increases the car's kinetic energy.

Because the speed increased, the kinetic energy increased. If you use a balance to weigh the fan car's mass, you might get a mass of 0.3 kg.

The fan car's initial KE would equal 0 J:

$$KE = ½ (m)(v^2) = ½ (0.3 \text{ kg})(0 \text{ m/s})^2 = 0 \text{ J}$$

As the fan car reaches 0.8 m/s, then its kinetic energy increases up to a maximum of almost 0.1 J:

$$KE = ½ (0.3 \text{ kg})(0.8 \text{ m/s})^2 = 0.096 \text{ J}$$

Remember—a large object moving slowly, like the train shown on the left, may have more kinetic energy than a small object moving quickly, like the race car, depending on the mass and speed (velocity).

Objects traveling vertically can have kinetic energy, too. Take a ball and drop it in free fall motion, like we discussed in Chapter 3. It accelerates under the influence of gravity, getting faster and faster. The ball has mass, and it has speed (velocity).

Because of its increasing speed, its kinetic energy is increasing also. Try it out for yourself.

Gravitational Potential Energy: The Energy of Position

Another type of energy in physics is GRAVITATIONAL POTENTIAL ENERGY (GPE). Imagine what happens when a weightlifter lifts a barbell off the ground.

What if the barbell fell out of the weightlifter's arms? It certainly wouldn't be good if it dropped on his toes! The barbell would be in a free fall motion and get faster and faster as it was pulled to the ground by the effects of gravity. So if the barbell gets faster, what happens to its kinetic energy? It increases because its speed increases.

Objects cannot gain energy without getting that energy from somewhere though. With your constant-speed car, the batteries provided the chemical energy that gave your vehicle its kinetic energy. This change from one type of energy to another is called ENERGY TRANSFORMATION.

What about the barbell? You could say the weight-lifter provided the energy to lift the barbell. But let's say the weightlifter lifted the barbell and placed it on a table.

When the barbell is on the table and at rest it has no speed, nor does it have any kinetic energy. What if the barbell accidently rolls off the table? Now it has speed and kinetic energy! But where did *that* kinetic energy come from?

When the weightlifter lifts the barbell against the effects of gravity and places it on the table, it has *gravitational potential energy (GPE)*. Gravitational potential energy is a type of energy that is stored in an object, in this case, the barbell, and is present based on its position. To gain GPE, you have to lift an object's weight (*mass or m × the acceleration due to gravity*, or *g*) above the ground (height, *h*). This means you can calculate GPE by taking an object's weight and multiplying it by its height above the ground (GPE = $m \times g \times h$—you'll often see this written as $mg \times h$). The ground is considered the REFERENCE LINE, or the BASELINE, and you measure the height from that line. At the reference line, *h* equals 0 meters.

Why call this energy "gravitational potential energy"? If the object falls, the potential is realized and the energy is transformed into kinetic energy. It was lifted up against gravity and now falls under the influence of gravity.

Let's do a few calculations. Take two objects with different masses. I used a 1-kg and a 2-kg mass. Start with both masses on the ground, the baseline or reference line. With height = 0, both masses have 0 Joules of GPE (GPE = $mg \times h$, $h = 0$ meters).

Now, let's lift both masses up to lab stool or other object.

By lifting the masses to the lab stool, you have given both masses GPE. It takes more effort to lift the 2-kg mass as it has twice the material inside. Both of the masses are the same height above the ground. You can measure this height with a meter stick. In this case, the lab stool is 0.6 meters above the ground. GPE = $mg \times h$, so you can calculate the GPE of the 1-kg mass as

$$GPE = 1 \text{ kg} \times 9.8 \text{ m/s}^2 \times 0.6 \text{ m}$$
$$= 5.88 \text{ Joules of energy}$$

The 2-kg mass has twice the GPE:

$$GPE = 2 \text{ kg} \times 9.8 \text{ m/s}^2 \times 0.6 \text{ m}$$
$$= 11.76 \text{ Joules of energy}$$

If the masses were to fall off the lab stool, you would have a loss of height and, consequently, a loss of GPE. But the energy is not lost, it just changes form. Ignoring the effects of friction and air resistance, you get the complete conversion of gravitational potential energy to kinetic energy:

$$mgh = \tfrac{1}{2}\, mv^2$$

Let's solve this equation for speed. For the 1-kg mass:

$$GPE = KE = mgh = \tfrac{1}{2}\, mv^2 = 5.88 \text{ J} = \tfrac{1}{2} \times (1 \text{ kg}) \times v^2$$

Solving this for *v* (velocity, or speed), you obtain a speed of 3.4 m/s. This calculation is an example of the LAW OF ENERGY CONSERVATION. Energy conservation is the change of forms of energy but not the change of amount of energy. Using energy conservation is powerful as it lets you figure out how fast an object is going without directly measuring its speed!

NOTE

The Law of Energy Conservation says that energy cannot be created or destroyed; it can only be transformed from one kind of energy to another.

In Chapter 3, you learned that all objects in free fall, dropped from the same height, achieve the same speed. This means that the 2-kg mass achieves the same speed as the 1-kg mass if it, too, falls from the lab stool. For the 2-kg mass:

$$GPE = KE = mgh = \tfrac{1}{2} mv^2 = 11.76 \text{ J}$$
$$= \tfrac{1}{2} \times (2 \text{ kg}) \times v^2$$

The speed (*v*) obtained = 3.4 m/s, the same speed as for the 1-kg mass! Your energy calculations confirm that all objects fall at the same rate in the absence of friction and air resistance.

Elastic Energy: The Energy of Elastic Materials

There is a third type of energy that is important in physics. Remember the catapult you built in Chapter 5?

The catapult initiates motion in the Ping-Pong ball. By pulling down on the lever arm, you trap a different type of potential energy in the rubber band or mousetrap spring. This potential energy is called ELASTIC ENERGY (E_{el}), and it represents energy trapped in the molecular bonds of the rubber band. My muscles provide energy to the rubber band or spring, which transfers this energy to the Ping-Pong ball as GPE and KE. Lots of energy transformations!

Experiment
Determine the Elastic Energy of a Rubber Band

Let's set up an experiment. You need the following items:

- A rubber band
- Meter stick
- Clamp
- Eyehook
- Support stick
- Spring scale

Clamp a support stick to the table, as shown in Figure 7-2. (I placed an eyehook in the support stick.) Line up the zero marking of the meter stick with the unstretched rubber band. Then stretch the rubber band a certain distance, 2 cm, for example, and measure the force reading from the spring scale. Replicate this for 4 cm, 6 cm, 8 cm, and so

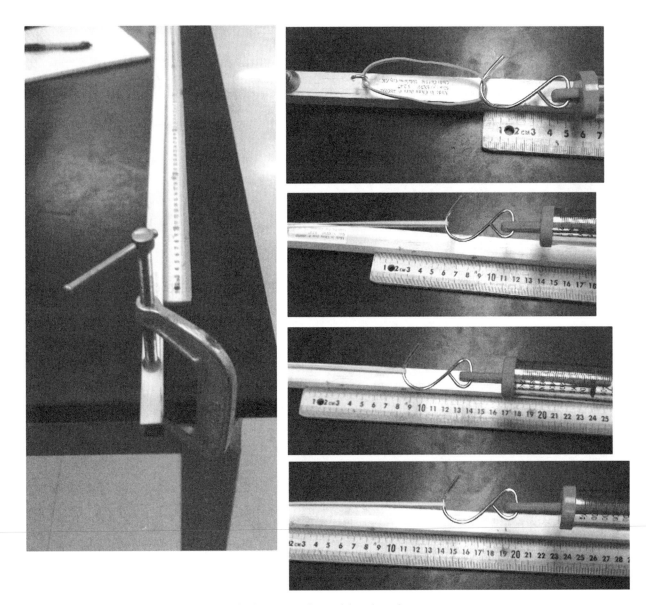

FIGURE 7-2 Measuring the forces needed to stretch a rubber band

on, until you have recorded a series of distances and the forces needed to stretch the rubber band those distances.

You can create a table for your rubber band data in Excel. You can see my data and graph in Figure 7-3. Graph the distance pulled or stretched on the x axis and the force needed to stretch the rubber band on the y axis. The pattern on your graph should be somewhat linear or diagonal in shape. Your data and graph, although different for your experiment, will hopefully look similar in shape.

This is a different type of graph than the distance-time graph and speed-time graph you created in previous chapters. With this type of graph, you're measuring distance and force. Instead of finding the slope of the line, as you did for the distance-time and speed-time graphs, you can take the area of the graph. When you know the area, you can find the elastic potential energy trapped in the rubber band.

To get the area, however, you need a graphing software program to help you obtain the area under the data points.

rubber band, force and distance		
distance pulled (cm)	distance pulled (m)	force (Newtons)
2	0.02	2.0
4	0.04	3.0
6	0.06	4.0
8	0.08	4.4
10	0.10	5.0
12	0.12	6.0
14	0.14	6.5
16	0.16	7.6
18	0.18	8.0
20	0.20	8.2
22	0.22	9.0
24	0.24	10.0

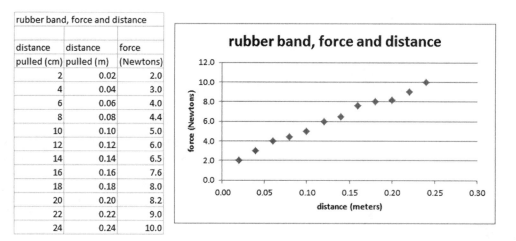

FIGURE 7-3 Data and graph for experiment to figure out the elasticity of a rubber band

NOTE

One graphing program that is free and found online is Graph (http://www.padowan.dk/graph/). Another graphing program, although not free, is Logger Pro (http://www.vernier.com/downloads/logger-pro-demo/).

For this experiment, download Graph and then import your Excel rubber band data into the program.

1. First, you need to develop a function for this data. The data appear linear, so apply a linear function by selecting the **Function** menu.

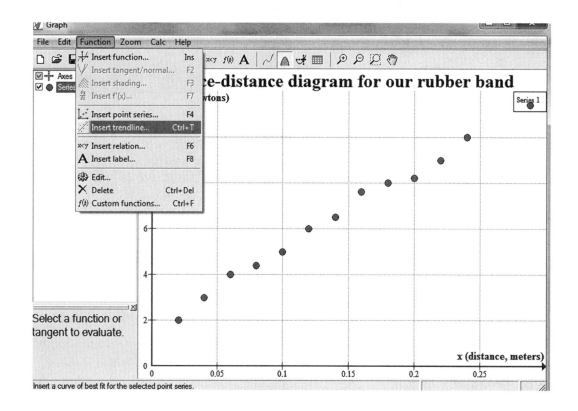

2. Select **Insert Trendline**, and when the Insert
   Trendline window appears, select **Linear**.

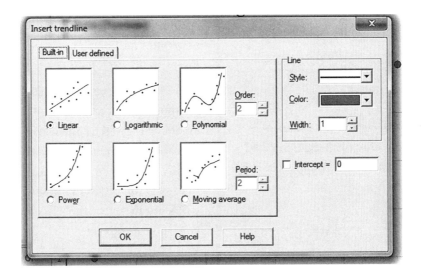

3. Click OK and you get a linear function, as
   shown here.

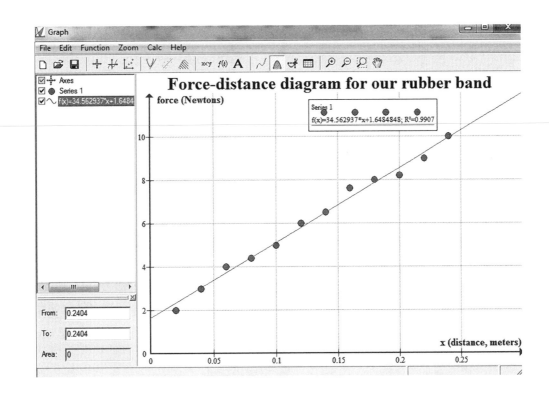

4. Navigate to the **Calc** menu and select **Area**.

5. Refer back to Figure 7-3 and the distance column from the Excel worksheet. Enter **0.02** for the From and **0.24** for the To input, as shown in Figure 7-4.

The area is 1.35 units of area. The area represents the elastic energy (E_{el}) trapped in the rubber band. If you had used this rubber band in your catapult, the Ping-Pong ball would have had the same amount of motional energy, ignoring all frictional losses.

Three types of energy are useful in studying the motion of objects: the energy of motion, or kinetic energy (KE); the energy of position, or gravitational potential energy (GPE); and the energy of elastic materials, or elastic energy (E_{el}). Energy is important in the study of physics. You will learn these concepts firsthand when building your mouse-trap car.

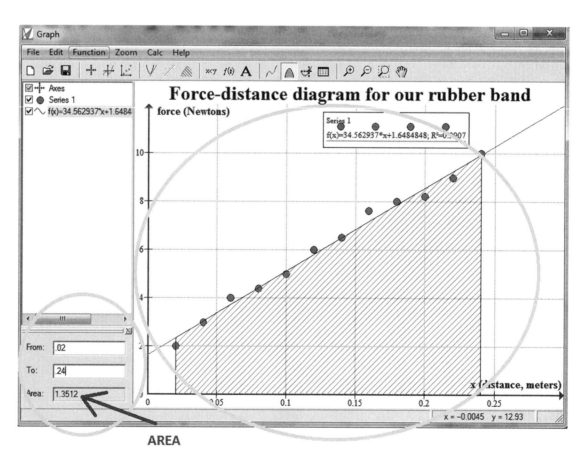

FIGURE 7-4 Calculating the energy trapped in the rubber band

Famous Scientists

The Joule (J), named for **James Joule** (1818-1889), is the unit of energy in physics and the metric system. But who was James Joule? He was born in England in 1818. As a young man, he was schooled at home and later tutored by the famous English chemist, John Dalton. He showed a keen interest in physics, chemistry, and mathematics. Joule later remarked that the time spent with Dalton greatly influenced his later pursuit of scientific interest.

But Joule's family owned a beer-making factory, and the family expected that the children would help run the business, so Joule and his brother ran the brewery. But James worked on experiments in the evening in his home laboratory.

In 1839, Joule began to study heat, electricity, and energy. He also began to write papers that he would present in scientific meetings. Many of his experiments dealt with the physics of *thermodynamics*, the study of heat and energy. At first, many scientists ignored his discoveries, but Joule continued his experiments until people realized he was making important discoveries.

In his now famous experiment involving the conservation of energy, Joule set up an experiment whereby a falling weight (*gravitational potential energy*) would spin a set of paddles in water. (A picture of his apparatus is shown here.) The paddles in the water stirred the water in the container and raised the temperature of the water by a set amount. By taking precise temperature readings, Joule was able to calculate the amount of energy that went into heating the water. This experiment changed how scientists thought about energy and heat and how energy can change forms but not be lost (*energy conservation*).

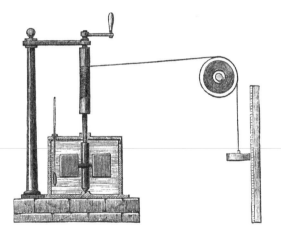

Project: Build a Mousetrap-Powered Car

A mousetrap-powered car (Figure 7-5) is a fun way to learn about energy conversion and transformation. The car moves by taking the elastic energy of a mousetrap and converting this energy to kinetic energy.

The basic design of the car is a body, which I made out of balsa wood, wheels from four CDs, corks to hold the wheels in place, axles to spin with the wheels, and a mousetrap that gives the car its energy to move. You can buy kits at hobby stores if you prefer not to build one from scratch. You can also find building plans on different websites. Some plans are designed to get the most distance from the mousetrap; others are designed for extreme speed.

FIGURE 7-5 A mousetrap-powered car

I am going to build a general-purpose car from scratch that you can modify if desired. You can use many different products and still build a successful car. For example, I've seen cars with cardboard or foam board bodies instead of balsawood bodies. Purchase supplies at a hobby and hardware store. As you walk through the steps to build the car, I will give you some alternatives.

Things You'll Need

Parts

- Two sheets of ³⁄₁₆" × 3" × 36" balsa (for the body of the car)
- Brass hollow tube, 12" × ³⁄₁₆" diameter (for the two axles), or wood dowels or other metal tubes would work, too
- Four CDs (for wheels). These are light, durable, easy to find, cheap (maybe free!), and each already has a small hole in the middle!
- Package of ten cork stoppers that fit the CD holes (to hold the axles to the wheels), or rubber stoppers or wine corks
- Four #8 washers
- One regular-sized VICTOR or equivalent mousetrap
- Brass or strong metal tube (hollow), 12" × ⅛" in diameter (for the lever or pulling arm)
- Paper clip
- Thread or string (Kevlar string, if possible)

Tools

- Drill
- Sandpaper
- Wood saw or hobby knife
- Meter stick or yard stick
- Vise or clamp
- Electric drill and drill bit set
- Hacksaw or saw to cut brass tube
- Markers/pens
- Pliers (slip joint or long nose)
- Duct tape roll
- Super Glue
- Hot glue sticks and a hot glue gun

BE CAREFUL!

Take care with the hobby knife or wood saw when you are cutting the balsa wood and be careful drilling with the electric drill. Use eye protection when using power tools. Always make sure an adult is available to supervise.

Steps to Build Your Mousetrap-Powered Car

1. Take one sheet of your balsa and cut it into equal halves with a wood saw or hobby knife. It is 36" long so you want two 18" inches pieces, measuring with your meter stick (or yard stick). These two pieces will be the sides of your car. Smooth the cut with some sandpaper. (As an alternative to balsa wood, you can use stiff cardboard or foam board. Make sure, however, that it is light and strong.)

A

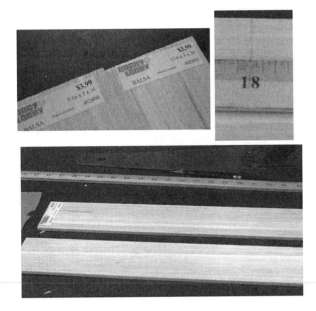

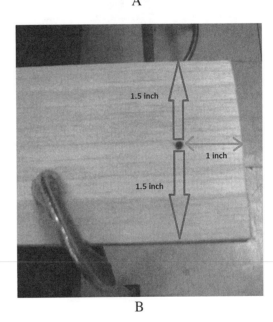

B

2. Clamp or tape these two 18-inch balsa pieces together (A). Clamping them makes it easier to drill the axle holes in the same location in both pieces.

3. After you have the two pieces clamped, measure 1" from the edge in the middle of the piece and make a mark (B). This is where you will drill the holes.

4. With the electric drill and a $\frac{5}{16}$" bit (or $\frac{1}{4}$" bit), drill a hole through both pieces of balsa. Sand the inside of the holes when finished.

5. Remove the clamp and turn the balsa around, clamping the other side. Drill a similar hole (1" from the edge) in the other side and sand to smooth them out.

6. You are going to create the axles from the 12" × $\frac{3}{16}$" diameter brass tubing. Use a hacksaw to cut the tube into two 6" pieces. The hollow tube might bend during the cutting; this is okay as you can bend it back into a straight shape with a pair of slip joint or long-nose pliers.

7. Push one cork stopper into each CD hole so it fits securely.

8. Mark a hole in the middle of each cork. Set the CD, with the stopper in place, on top of a roll of duct tape and tape the CD onto the roll so you can drill without it moving around.

9. Using a 1/8-drill bit, drill a hole in each cork. Don't drill through the cork stopper as this might split the stopper. You just need a small pilot hole to push the axle in.

10. Push one end of the axle into the hole created in the cork stopper. Do this for both axles. Push the axle in carefully but securely.

11. Now you'll build the top and bottom of the car. From the other pieces of the balsa wood, cut two 12" pieces with the wood saw and sand smooth. One of these pieces is the top and the other is the bottom.

12. Line up the axle holes in the side pieces with two pens, as shown in the top photo in Figure 7-6. This keeps the axle holes in the same position as you glue on the top and bottom pieces. The two side pieces are 18" long and the top and bottom pieces are 12" long.

13. With a meter stick, measure 3" from one side, as shown in the second photo in Figure 7-6. Glue the top on with the hot glue gun. After the glue sets, turn the body upside down and glue the bottom piece.

 BE CAREFUL!

Hot glue guns are hot and can cause serious burns. Use care and make sure an adult is present to supervise.

14. Now you're ready to mount the wheels and axles to the body, as shown in Figure 7-7. Put a washer on the axle with one CD/cork stopper and put this into the axle holes and then add another washer and attach the axle to another CD/cork stopper, as shown in the figure. Do this for both sets of wheels. Push the axles snugly into the cork. The car should roll. Give it a push and make sure it rolls straight. Adjust the CDs on the cork stoppers if needed.

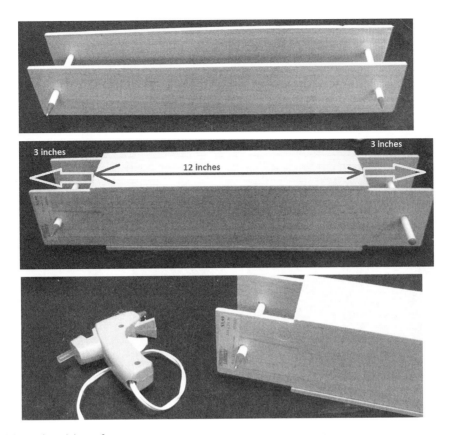

FIGURE 7-6 Building the sides of your car

FIGURE 7-7 Mounting the wheels and axles on the car

15. Now you're ready to prepare the mousetrap. First, you need to remove several things from the mousetrap.

   a. Remove the part that holds the "cheese" on the mousetrap with a pair of slip joint or long-nose pliers. Just pull it out. You can discard this part.

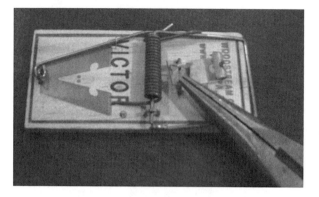

   b. Remove the metal loop that holds the mousetrap locking (setting) bar into place. You can discard this loop.

   c. Another loop holds this metal locking bar to the edge of the mousetrap. Using the pliers, remove this loop, too. You can discard the metal loop but save the locking bar.

   d. Cut the snapping arm at the position indicated on the next photo. Use a pair of long-nose pliers or wire cutters. Save this metal snapping arm.

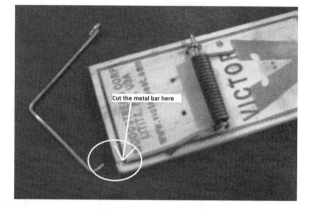

Cut the metal bar here

Here is everything you've removed from the mousetrap.

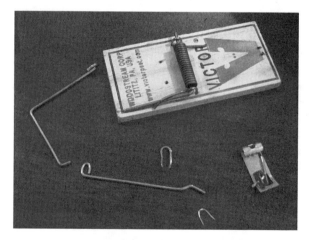

16. Take the locking bar and straighten the non-loop end with your pliers.

17. Slide this straightened part of the locking bar into the lever arm. The lever arm is the 12" × ⅛" brass or metal hollow tube. Slide the locking bar all the way into the hollow lever arm and use Super Glue to fasten securely.

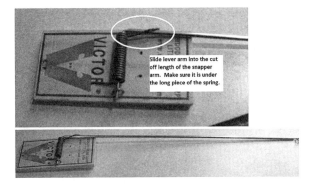

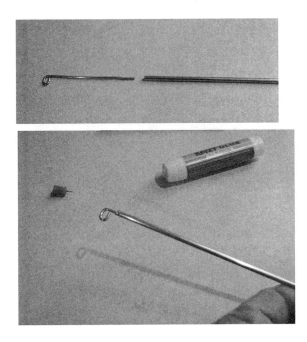

18. You'll attach the other end of the lever arm to the mousetrap. Slide the lever arm into snapping arm that you cut off. Make sure the lever arm slides into the snapper arm but under the long piece of the mousetrap's spring, as shown next.

The lever arm should be snugly placed and attached to the mousetrap. You can rotate all as one unit.

19. Take a paper clip and unwind it. Using your pliers, create a loop from the unwound paper clip and attach it around one of your axles. This axle will be your back or power axle. See Figure 7-8.

20. With your pliers, twist the looped paper clip around the axle so it is snug around the middle part of the axle.

21. Attach this loop with hot glue or Super Glue (Figure 7-8).

22. Finally, cut off the excess paper clip loop but leave a small segment about ¼- to ½-inch long. You'll use this axle hook to attach the string. (You could also possibly use a twisty-tie or zip lock to create this axle hook.)

23. Earlier, you cut and removed the snapper arm. You'll use this piece on the mousetrap car to hold the lever arm when you wind up the string on the back axle. Take the snapper arm and hot glue the loop of the snapper arm to the side of the car, about an inch from the edge of the top of the car, as shown in Figure 7-9.

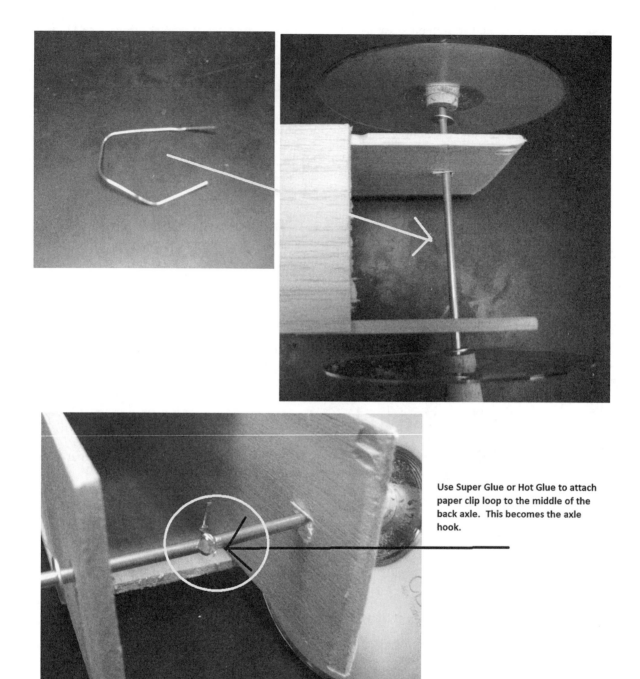

Use Super Glue or Hot Glue to attach paper clip loop to the middle of the back axle. This becomes the axle hook.

FIGURE 7-8 Take a paper clip, unwind it, and create a loop around one of your axles.

The top of the snapper arm should be about a ½ to ¾ inch above the top of the car (in other words, there should be a gap of about a ½ to ¾ of an inch from the top of the car to the snapper arm).

24. Now you're ready to attach the mousetrap to the top of the car's body. Pull the lever arm back, activating the mousetrap as shown in Figure 7-10.

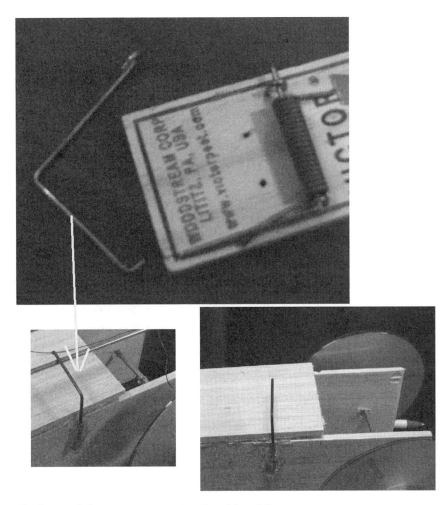

FIGURE 7-9 Glue the loop of the snapper arm to the side of the car.

25. Align the mousetrap on the top of the car so the metal loop of the lever arm is directly above the axle hook on the back axle of the car. As you can see in Figure 7-10, I made a mark on the top balsa piece with a pen to indicate the correct placement of the front edge of the mousetrap. Then using hot glue, attach the mousetrap to the top of the car.

26. Now you're ready to attach the string to the lever arm. You can use string or thread. Securely tie one end of your string to the metal loop of the lever arm, as shown here.

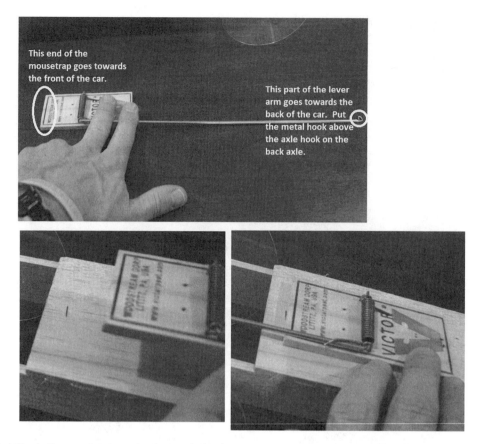

This end of the
mousetrap goes towards
the front of the car.

This part of the lever
arm goes towards the
back of the car. Put
the metal hook above
the axle hook on the
back axle.

FIGURE 7-10 Attach the mousetrap to the car's body.

27. Measure out a length of the string from the lever arm loop to the back axle. Make a secure loop here. Make sure this loop just reaches the axle hook of the back axle and no longer.

28. Now you're ready to wind up the string. Rotate the lever arm toward the back of the car and hook the lever arm under the snapper arm to hook the lever arm back while you wind the string around the back axle (Figure 7-11).

29. Attach the loop on the end of the string to the axle hook on the back axle. Rotate the wheels counterclockwise while you hold the string on the axle hook.

 **HINT!**

Keep your windings tight and keep winding it around the back axle until you have wound the complete string. Don't get any of your windings too close to the edge of the sides of the car. I have a short movie of me winding my car on the website.

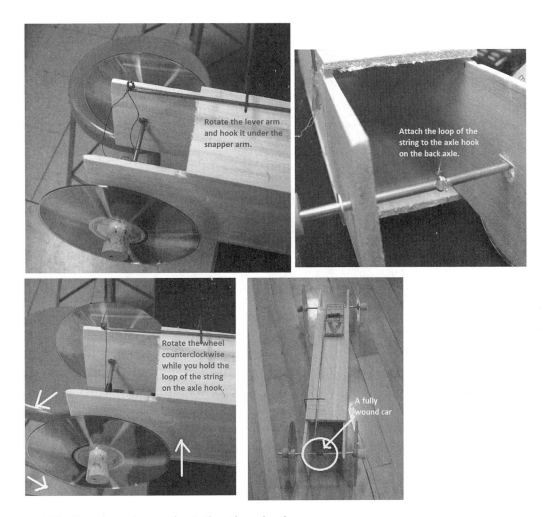

FIGURE 7-11 Winding the string and rotating the wheels

Once the car is wound up, you are ready to give it a test run. You need a long flat area for your test. I tested mine in a gym. When you are ready, carefully hold the car and lift the lever arm over the snapper arm. The lever arm should be pulled slowly around by the energized mousetrap spring. As the lever arm is pulled back toward the front of the car, it should gain speed and get faster. Your car may turn some. To correct for this, you can adjust your wheels on the cork stoppers. There may be several other minor adjustments you'll need to make. A well-built car can travel over 15 meters. I've posted several videos on the book's website of my car moving across the gym, as shown in Figure 7-12.

FIGURE 7-12 My test run of the mousetrap-powered car

Summary

Building a mousetrap-car is a great project for this chapter as the vehicle "works" by transforming energy. We activate the mousetrap, pulling back on the lever arm to give the mousetrap elastic energy. This energy transforms slowly into kinetic energy, moving the car along the gym floor. Friction and air resistance dissipate this kinetic energy as the car moves, and this represents an energy transformation, too! Great job on your mousetrap car!

CHAPTER 8

Whacks and Bangs: The Physics of Collisions

IN THE PREVIOUS CHAPTER, I DISCUSSED KINETIC energy, gravitational potential energy, and elastic energy. Kinetic energy is the energy of motion. Gravitational potential energy is the energy an object has because of its height above the ground. Elastic energy (or elastic potential energy) is energy stored in an object due to shape distortion, such as a stretched rubber band or a compressed spring.

When energy is conserved, energy can change forms but is not lost or gained. For example, a runner takes food energy and transforms this into energy of motion and/or energy of height (Figure 8-1).

A

B

FIGURE 8-1 The runner running on a flat track (A) has kinetic energy or energy of motion. When the runner is running up a set of stairs (B), the runner has both kinetic energy and is gaining gravitational potential energy because of a gain of height above the ground.

Energy conservation is an important idea, not only in physics but also in many other sciences as well.

A concept similar to kinetic energy is called MOMENTUM. Momentum combines two concepts: mass and speed (or velocity). As discussed in Chapter 4, mass is the matter that comprises an object—its atoms. In physics, you measure mass in the metric unit called the kilogram. For example, a runner might have a mass of 70 kilograms. Speed is typically measured in meters per second. If the 70-kilogram runner is running 3 meters per second, then the runner has a momentum of 70 kg × 3 m/s = 210 kg*m/s. A runner with more mass or more speed would have greater momentum.

HINT!

For an object to have momentum, it must have both speed and mass.

In this chapter, you'll learn about momentum and how it can be moved from one object to another in different types of interactions. You'll also investigate the momentum of a rocket engine and how this momentum gives the rocket its upward motion.

The Conservation of Momentum and Collisions

Both kinetic energy and momentum involve mass and speed. The amount of energy in an object can change form, but the total amount of energy doesn't change. A similar theme is seen with momentum. Momentum is conserved, too. There is a difference, however, between the conservation of energy and the CONSERVATION OF MOMENTUM. With conservation of energy, changes take place in a single object. With conservation of momentum, two or more separate objects come together and collide in some way. In these COLLISIONS, the objects exchange and transfer momentum. If you ignore friction and any other resistive forces, the momentum the objects had before the collision is equal numerically to the momentum the objects have after the collision.

Types of Collisions

Many types of collisions occur on our planet and throughout the universe, and you can use momentum and its conservation to understand these collisions. There's the collision of cars in a fender-bender; the collision of asteroids, moons, or planets in our solar system; and the collision of atoms and subatomic particles in particle accelerators like Fermilab or CERN. Collisions exist among the smallest things in our universe all the way to the largest. There are also collisions in sports. Two football players collide during a tackle or a tennis racquet and the ball collide when the player swings the racket, hitting the ball. Collisions involve forces, and these forces act to exchange momentum. And in all collisions, momentum is conserved if you can ignore friction and resistive forces. Then the momentum of the objects before the collision is equal to the momentum after the collision.

Physicists have designed experiments where particles collide. In Figure 8-2, you see the beam tube at the Relativistic Heavy Ion Collider Tunnel at the Brookhaven National Lab. Atoms are accelerated to close to the speed of light in these beam tubes and then brought together in collisions that help physicists learn about the building blocks of the universe.

Collisions are classified into two major categories: ELASTIC and INELASTIC. Elastic collisions are BOUNCY. Two objects hit and bounce right off of each other. In a perfectly elastic or bouncy collision, both momentum and kinetic energy are conserved. No heat or sound waves are generated

FIGURE 8-2 High-speed atomic collisions take place in the Relativistic Heavy Ion Collider. (Used by permission of Brookhaven National Lab.)

during the collision or any other process to dissipate any energy. Perfectly elastic collisions are rare in nature, existing in only some subatomic collisions and gravitational interactions. In most everyday collisions, called MACROSCOPIC interactions, some heat and friction are generated and sound waves are formed. These processes dissipate energy and keep the collision from being perfectly elastic. Some collisions, however, are approximately or *almost* perfectly elastic. For example, billiard balls are very hard and lose little energy when two balls hit and bounce off of each other. These collisions are close to perfectly elastic, which is why physics textbooks often use billiard balls as examples of elastic collisions. A little heat and sound is produced, which acts to dissipate some of the kinetic energy.

The other type of collisions are perfectly inelastic, or STICKY collisions, and act when two objects collide and stick together. After the collision, the stuck-together objects move away at some common speed. Have you ever seen two cars whose fenders are jammed together after a collision? This is an example of an inelastic collision as the two objects are now stuck together. Or what about an outfielder catching a fly ball in his or her mitt? This is an inelastic collision, too, as the glove and the fly ball become one entity.

INTERESTING FACT
RECOIL

There is another type of collision that is not really a collision at all, but it does represent an interaction between two objects. This type of collision is called a RECOIL reaction. Two objects are together as one unit and then split apart. The word *recoil* is often used when referring to shooting a gun. The bullet and the gun make one unit. When fired, the bullet goes one direction and the gun pushes back in the other direction. The gun shooter feels this push-back—or recoil—in his or her shoulder. The bullet has momentum in one direction, and the gun has momentum in the other direction.

A famous simulated recoil situation discussed in many physics books involves an astronaut stranded in space. The astronaut needs to get back to the spaceship but the tether has broken and the astronaut has no way back. The solution to the situation involves the astronaut throwing something forward like a tool. The tool goes forward and the astronaut is recoiled backward, toward the spacecraft.

The Impulse-Momentum Equation

The product of *mass* and *speed* (or *velocity*) is *momentum*. And the conservation of momentum is when the total momentum before a collision equals the total momentum after the collision. Another important momentum equation comes directly from Newton's second law of motion, which you learned about in Chapter 6.

Newton's second law of motion is $F_{net} = m(a)$. The *net force* on an object is equal to the object's *mass*

times its *acceleration*. How can you convert this to momentum? Acceleration is a change in speed (either positive or negative), so *acceleration* is equal to a *final speed* minus an *initial speed* divided by *time:*

$$a = \frac{v_{final} - v_{initial}}{time}$$

In equation form, you can write it this way:

$$F_{net} = m(a)$$

or

$$F_{net} = m\left(\frac{v_{final} - v_{initial}}{time}\right)$$

Then you can move *time* to the other side of the equation:

$$F_{net}\,(time) = F_{net}t = m(v_{final}) - m(v_{initial})$$

$m \times v$ is the equation for momentum. Let's rewrite this one more time:

$$F_{net}t = final\ momentum - initial\ momentum$$

There you have it: the momentum equation comes from Newton's ideas. In words you might say, "I can change an object's momentum if I apply a force on that object for some amount of time." The $F_{net}t$ part of the equation is given a special name—IMPULSE. This whole equation is called the *impulse-momentum equation* or *theorem*. It's an easy equation to write but it expresses a very powerful idea.

Using the Impulse-Momentum Equation in the Real World

Let's say your car is stopped at a red light. The light turns green. You now want to move your car and get it up to speed. The speed limit is 50 miles per hour. The car has a mass, for our example, 1000 units of mass. So you want to take this mass from a speed of 0 at the red light to 50 miles per hour when the light turns green. This is a change of momentum: your final momentum would be the product of *mass* and *speed,* which equals 50 × 1000 or 50,000 units of momentum minus your initial momentum (which would be 0 as your speed is 0 miles per hour initially).

The change of momentum would be 50,000 units of momentum. The impulse-momentum equation states that this momentum change can be done by a force acting through time, but the equation doesn't state the size of each force or the length of time. You might have a large force acting through a small period of time or a small force acting through a large period of time. In both cases, you can achieve 50,000 units of momentum:

$$F_t = 50,000 \text{ or } Ft = 50,000$$

So you could have a force of 50,000 units acting for 1 second or a force of 25,000 units acting for 2 seconds or even a force of 10,000 units acting for 5 seconds. Any of these combinations when multiplied gives you the exact same 50,000 unit change of momentum. Going back to the car, you could apply a large force on the accelerator (press down hard) and get to 50 mph quickly or apply a small force (press more slowly) for a longer time period. Either way, you can get your car to 50 mph.

Automobile engineers use the impulse-momentum equation to save lives! Let's say that I accidently drive into a tree, resulting in a rapid (large force, small time) change of momentum for my car. But remember what I talked about in the last chapter— I'm moving with my car and my momentum needs to change, too. If I am not strapped in with a seatbelt or protected by an airbag, my head and body might continue moving forward until I am stopped by the windshield, which acts to change my momentum rapidly, too (large force, small time).

This large force on my head and body can cause serious injury. Automobile engineers have designed and installed airbags in cars to manipulate the force and the time part of the momentum-impulse equation. So if I hit the tree, the car's airbags will deploy (see Figure 8-3). If so, my head hits the airbag instead of the windshield. My momentum is changed the same amount as would happen when I hit the windshield but the force is smaller while the time is greater. Same impulse but now instead of BIG FORCE, small time, I have small force, LARGE TIME. The wreck is now more survivable.

FIGURE 8-3 Airbags save lives by applying a small force over a large time.

Famous Scientists

Dr. John Paul Stapp was a medical doctor and Air Force aerospace researcher. He grew up in Texas where he studied English and zoology at Baylor University. He later received his doctorate in biophysics along with a medical degree. In 1944, Stapp joined the Medical Corps of the United States Air Force and studied the effects on the human body of high altitude airplane flights. As a result, he spent many hours in unpressurized airplane cabins and cockpits at altitudes up to 45,000 feet!

In the late 1940s and 1950s, Stapp began studying excessive changes of momentum during rapid accelerations and decelerations and how to help the human body survive these rapid changes. To conduct these experiments, Stapp strapped himself into a rocket sled. A rocket sled uses the thrust of rocket engines to provide these extreme forces to change motion quickly. In 1954, Stapp "piloted" the *Sonic Wind 1,* shown here, a rocket sled that slowed him down from 632 miles per hour to rest in a mere 1.4 seconds. In the second photo, you can see three time-lapse pictures of Stapp during an extreme deceleration of the rocket sled. Using the research from Dr. Stapp and others, special harnesses and ejection seats were invented to help pilots survive high-speed aircraft malfunctions.

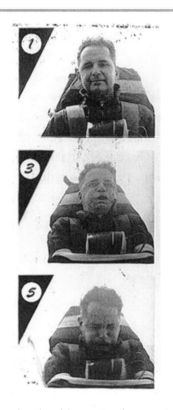

While conducting his research on extreme acceleration and deceleration, Stapp began to study automobile safety as well by putting crash test dummies into cars and projecting them into barriers. Thus, began the idea of manipulating impulse (a smaller force over a longer time) to make cars safer in the event of an accident. Soon volunteers were testing harnesses for cars that would eventually be called "seatbelts."

Dr. Stapp's research led to many other safety devices on cars, including padded dashboards, steering wheels, and bumpers. Today we have front and side airbags, crumple zones on cars, and car seats. Many of these devices help save lives by manipulating impulse to create smaller forces that act through more time.

Impulse and Rockets

Rocket scientists and engineers have to know quite a bit about the impulse-momentum equation. The force given to rockets is called THRUST. Thrust is the force that moves the rocket through the air. In a rocket, fuel combusts with an OXIDIZER (source of oxygen). In this chemical reaction, hot gas is accelerated and exhausted from the rear of the rocket. The rocket itself is then accelerated upward due to Newton's third law of motion. For every action, there is an equal and opposite reaction. This is also considered a recoil reaction. In terms of impulse-momentum, an impulse (*force* multiplied by *time*) is created by the chemical reaction, and this impulse creates a change of momentum for the rocket.

The Model Rocket Engine

You're almost ready to take this equation and apply it to model rockets. Small model rocket engines use physics principles that are similar to those used by large rockets. Figure 8-4 shows you the anatomy of a model rocket engine. Various models of engines provide different amounts of impulse. A larger impulse engine can launch a heavier rocket and/or achieve a higher altitude. A typical model rocket engine is two to four inches long, with a diameter of maybe one-half to one inch. You place an igniter in the nozzle and activate the rocket electrically. The propellant burns, providing thrust. This burn might take up to one or two seconds. The product of the force (thrust) and the time provides the impulse. The delay charge may offer a smoke

trail for tracking the rocket as it coasts upward (modeling free fall in an upward direction). Then the ejection charge provides a push upward against the rocket's nose cone forcing the parachute for rocket recovery.

Engine Coding and Symbols

Model rocket engines are given a code based on the impulse delivered by the engine. The National Association of Rocketry (http://www.nar.org/index .html) set up this code. The next table shows a partial listing. The first letter represents the impulse delivered by the engine. Notice the impulse is doubled from one letter to the next.

First Letter	Impulse Delivered (Newton-sec)
A	1.26-2.50
B	2.51-5.00
C	5.01-10.00
D	10.01-20.00

The first number after the letter represents the average thrust (force) of the engine. For example, you can buy a B6-6 rocket engine and C6-3 rocket engine. Both of these engines have the same average thrust (6). The C engine would have more impulse, however, and would fire for a longer time period. The number after the dash is the time from the propellant burnout until the ejection charge that forces the parachute out. Some engines have smaller impulses. These are often given as ¼ or ½ in front of the first letter *A*. For example, an engine coded ½A3-2T would have half the impulse of an A engine. Here you can see a photo of two different engines: an A3-4T and a B4-4.

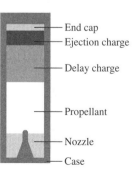

FIGURE 8-4 Anatomy of a model rocket engine

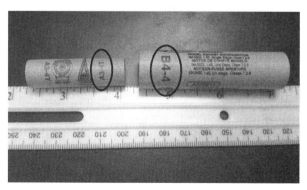

Project: Build an Engine Holder for the Digital-Balance Method

You can test the impulses in model rocket engines using a couple of methods. The first gives good results and can be done with equipment purchased from your local hobby, hardware, or department store along with a video camera and computer. The second method requires more expensive equipment; I'll describe it toward the end of the chapter.

Things You'll Need

Parts

- PVC fitting that is 3" long, with a 2 or 2.5" inside diameter, and a wall thickness of around ¼".

- Four 2" ¼-20 bolts, which you'll use to securely attach the engine to the PVC tube.

- Digital balance. You can buy inexpensive digital scales online or at discount stores. Make sure your scale can record mass in grams (at least up to 1000 grams or 1 kilogram).

- Rocket engines. I suggest getting a couple of different sizes; I used A3-4T and B4-4 engines. Each packet usually contains three or four engines. The engines should come with several igniters, which you will need.

- Model rocket launcher ignition system.

NOTE

In the next chapter, you'll build an ignition system from parts. Feel free to jump ahead and build one to use here.

Tools

- Drill with a ³⁄₁₆" drill bit
- Tap for the bolts
- Eye protection
- Ruler
- Protractor

BE CAREFUL!

Work with an adult when using power tools and when igniting model rocket engines. Use safety gear when igniting engines (eye protection and ear protection, too, and have a fire extinguisher handy).

Let's build the engine holder:

1. With the ³⁄₁₆" drill bit, drill four holes in the PVC fitting. Use the protractor to make sure the holes are 90 degrees apart and drill about ½" from one of the ends of the PVC fitting, as shown here.

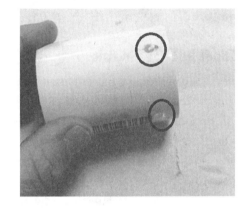

2. Put the four bolts in the holes you've drilled in the PVC fitting.

3. Place one of the rocket engines in the center of the fitting and secure it tightly with the four bolts, as shown here.

BE CAREFUL!

Now is a good time to put on your safety glasses or goggles!

4. Insert an igniter into the engine and add a plug. Push the end of the igniter into the nozzle of the engine and push the plug down on top of the igniter.

5. Take the two end wires of the igniter and spread them away from each other. Attach the two alligator clips from the model rocket engine launcher to these two end igniter wires.

Now you're ready to take your engine and holder outside to find a flat concrete surface for your scale and engine bracket.

BE CAREFUL!

Make sure the place you select is away from people, obstacles, and flammable substances. The recommended distance you should stand away after you ignite an engine is 15 feet.

Experiment
Determine the Model Rocket Engine's Impulse Using the Digital-Balance Method

Take your digital scale and place a sheet of cardboard or thick set of papers between the engine holder and the top of the scale, as shown in Figure 8-5, to protect it from the burning engine.

Place the video camera on a tripod (or otherwise secure it to hold it steady) and point the camera down so you can film the numbers on the digital scale. Make sure the camera is away from the engine exhaust.

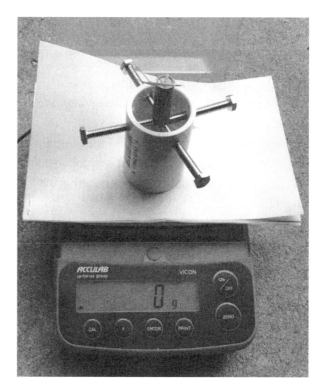

FIGURE 8-5 Setting up the engine holder on the digital scale

Make sure the model engine launcher ignition system is set up with the alligator clips on the two igniter wires and that it is ready to fire. Zero out the scale before you begin filming, as in Figure 8-5.

BE CAREFUL!

Make sure you are far enough away from the engine with your safety gear on, a fire extinguisher handy, and an adult present.

Fire the engine when you're ready. You'll notice the igniter and wires get pushed away from the nozzle as the engine is ignited.

BE CAREFUL!

If there is a problem and the engine does not ignite, approach the engine carefully and slowly, keeping your face away from the engine itself. Disconnect the alligator clips and then check the igniter placement and plug.

As the engine fires, the upward thrust pushes the engine holder down and the video camera records how the digital scale numbers change (see Figure 8-6). For a small engine (for example, the A3-4T that I tested), firing is complete within a few seconds. After the firing is complete, approach the engine slowly and carefully. Turn off the camera. The spent engine will be hot; allow it to cool before handling. Your data should be safely stored in the camera.

Now you'll use the video to collect your data for graphing. The goal is to get time and force data in order to graph these two variables and calculate impulse. The area of your graph will give you the impulse, and you can check to see how close your impulse compares with published results.

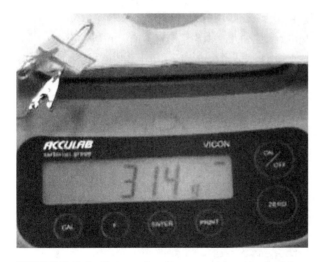

FIGURE 8-6 As the engine fires, pushing the holder down, the numbers on the digital scale change.

Thrust, Time, and Impulse Graphs

Model rocket engines burn propellant in a special way to produce the thrust. You can diagram this on a FORCE-TIME graph. Time is on the x axis and force is on the y axis. In the graph shown here, notice that at first, the force peaks. This peak represents the initial firing of the rocket engine. As the rocket engine fires, the thrust decreases to a stable, constant value. Depending on the engine used, this stable thrust might last for up to one or more seconds.

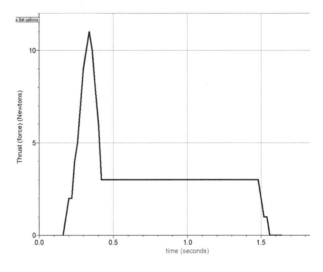

Remember, *impulse* is the product of *force* multiplied by *time*. On the graph you can show this product by taking the area under the line, as shown here. The larger the impulse, the greater the area and the greater the change of momentum for the rocket. For a given rocket, a larger impulse gives the rocket greater acceleration and greater altitude.

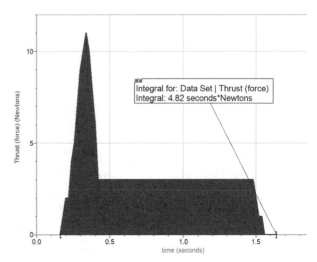

Create Your Impulse Graphs

Because the readings on the digital scale change so quickly, you have to frame the video in Tracker Video Analysis to see the changes. Let's create the impulse graph for your engine:

1. Upload the video into Tracker to analyze the frames. Find the first frame in the video where the digital balance reads the start of the engine firing. For my video (small_engine_on_digital_scale), which you can download from the website, the scale starts reading at frame 89. At frame 88, the igniter has blown off and the scale reads 0, and at frame 89, it reads 4 grams, so I make frame 88, my 0 data point, and frame 89, my 1 data point.

2. Step through the video with the Tracker software frame by frame. Note the frame number, the data point number, and the reading of the digital balance as you step through the video. This will be the data for your graph. Your last data point is when the engine has completed its firing (when the scale either returns to 0 or doesn't change for a while). With my video, the scale stopped changing at frame 154, so I knew the engine had completed its firing. I did place my video on the website for the book if you would like to view it.

3. Create an Excel file. Make three columns: the first for the frame number, the second for the data point number, and the third for the scale reading. For my video I took data from frames 88–154, which gave me 67 data points (0–66). You can see a screenshot of part of my Excel file next.

small_engine_data_table_on_excel

	A	B	C	D
1	Estes Model Rocket Engine A3-4T			
2	Data using digital balance and rocket engine holder			
3				
4	frame no	data point no.	scale reading (grams)	
5	88	0	0	
6	89	1	4	
7	90	2	4	
8	91	3	4	
9	92	4	4	
10	93	5	4	
11	94	6	4	
12	95	7	314	
13	96	8	314	
14	97	9	314	
15	98	10	314	
16	99	11	314	
17	100	12	314	
18	101	13	210	

4. Convert your data points into a time. Call data point 0, 0 seconds. Because you're using digital video, you can approximate the time between one frame and the next to be 30 frames per second. So the time between one frame and the next is $1/30^{th}$ of a second; in decimals, this would be 0.033333. Your second data point occurs at 0.033333 seconds, and your next data point is twice this amount, or 0.066666.

   a. Create a new column named **time (seconds)** in Excel.

   b. Take each data point and multiply it by 0.033333 seconds. (My last, or 67th, data point is approximately 2.2 seconds.)

5. Convert your scale reading from grams to kilograms. Create a new column called **scale reading (kilograms)** and divide the mass in grams by 1000 as there are 1000 grams in 1 kilogram. On the next page, you can see my Excel spreadsheet after completing Steps 4 and 5.

small_engine_data_table_on_excel

	A	B	C	D	E
1	Estes Model Rocket Engine A3-4T				
2	Data using digital balance and rocket engine holder				
3					
4	frame no	data point no.	scale reading (grams)	time (sec)	scale reading (kilograms)
5	88	0	0	0	0
6	89	1	4	0.033333	0.004
7	90	2	4	0.066666	0.004
8	91	3	4	0.099999	0.004
9	92	4	4	0.133332	0.004
10	93	5	4	0.166665	0.004
11	94	6	4	0.199998	0.004
12	95	7	314	0.233331	0.314
13	96	8	314	0.266664	0.314
14	97	9	314	0.299997	0.314
15	98	10	314	0.33333	0.314
16	99	11	314	0.366663	0.314
17	100	12	314	0.399996	0.314
18	101	13	210	0.433329	0.21

6. Now you want to convert mass in kilograms to a force (thrust). Because the rocket creates an upward thrust, the downward force pushing on the scale is a weight. Multiply the mass in kilograms by little g, which you know from earlier chapters to be 9.8. Basically, you're using an equation to find the weight of objects ($W = mass$ in kilograms multiplied by little g, and little g on Earth is 9.8 m/s/s) in Newtons, which will be your thrust. Create a new column named **Thrust (Newtons)**, and set up an Excel formula to help you do the math. Figure 8-7 shows my complete Excel file.

NOTE

If you're not sure about setting up a formula in Excel, ask an adult or use the Excel Help.

small_engine_data_table_on_excel

	A	B	C	D	E	F
1	Estes Model Rocket Engine A3-4T					
2	Data using digital balance and rocket engine holder					
3						
4	frame no	data point no.	scale reading (grams)	time (sec)	scale reading (kilograms)	Thrust (Newtons)
5	88	0	0	0	0	0
6	89	1	4	0.033333	0.004	0.0392
7	90	2	4	0.066666	0.004	0.0392
8	91	3	4	0.099999	0.004	0.0392
9	92	4	4	0.133332	0.004	0.0392
10	93	5	4	0.166665	0.004	0.0392
11	94	6	4	0.199998	0.004	0.0392
12	95	7	314	0.233331	0.314	3.0772
13	96	8	314	0.266664	0.314	3.0772
14	97	9	314	0.299997	0.314	3.0772
15	98	10	314	0.33333	0.314	3.0772
16	99	11	314	0.366663	0.314	3.0772
17	100	12	314	0.399996	0.314	3.0772
18	101	13	210	0.433329	0.21	2.058

FIGURE 8-7 My Excel file with all the data for graphing in place

7. Excel isn't set up to find area easily so you need to paste the Time and Thrust columns into a graphing program. Logger Pro, a software program used by many high schools and colleges for graphing, is available from Vernier Software and Technology (www.vernier.com). You can download the 30-day trial and/or buy it (http://www.vernier.com/downloads/logger-pro-demo/). The other option, which I explain in Step 9, is to use a free program called Graph (http://www.padowan.dk/graph/).

8. In Logger Pro (I use version 3.8.4), paste the two data columns from Excel: the Time data in the *x* column and the Thrust data to the *y* column. You may need to renumber your axes to enlarge the graph. Now click the **Area** menu button.

   The impulse reading is given in Newton-second units, as you can see in Figure 8-8. The digital balance method gives me a reading of 2.26 Newton-seconds.

9. In Graph, paste the Time and Thrust columns from Excel into the Graph program. Select the **Function** command and then select **Insert**

point series… Put the Time data in the *x* data column and the thrust data in the *y* data column. Feel free to modify the title and axes by choosing the **Axes** command from the **Edit** menu. Click the **Area** button, which is circled in Figure 8-9.

10. To get the area, you designate a function. Choose **Function** from the menu bar, and then select **Insert Trendline…** Select **Moving Average** and click **OK**. You should see the area the lower left of the Graph window (Figure 8-10). The Graph program gives me an area or impulse of 2.2575 Newtons-seconds, which is close to the value obtained in Step 8 with Logger Pro.

You can repeat this experiment using a different size rocket engine. I tested the B4-4 engine and posted the video on the website (larger_engine_ showing_digital_scale.). I replicated the steps in this section to access my data, create the force-time graph, and calculate the area (impulse) of the graph. When I finished, I got an impulse of 4.016 Newton*seconds. What impulse did you get with your larger model rocket engine?

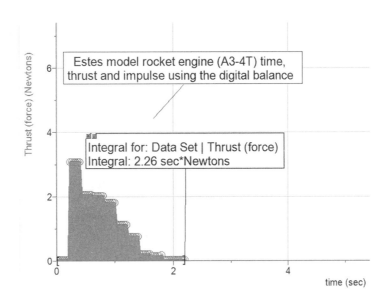

FIGURE 8-8 The area of the force-time graph for the A3-4T engine in Logger Pro. The area represents the impulse delivered by the engine to the digital balance.

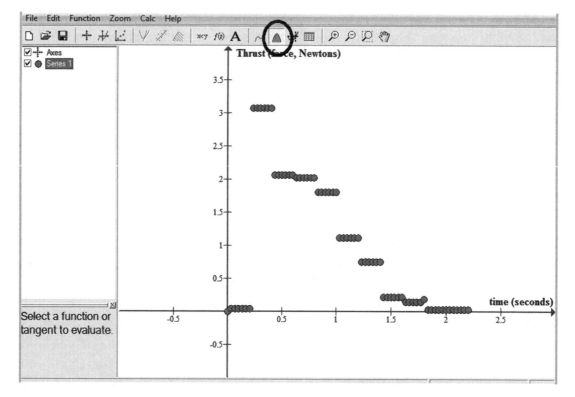

FIGURE 8-9 The force-time graph using the Graph program

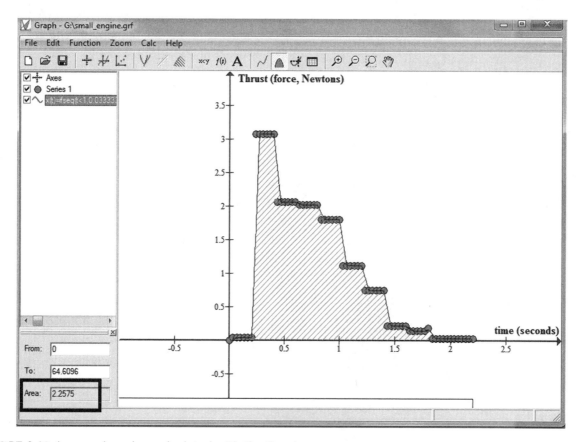

FIGURE 8-10 Area, or impulse, calculated with the Graph program

Experiment
Determine the Model Rocket Engine's Impulse Using the Force-Probe Method

The second method for getting the model rocket engine's impulse might be hard to replicate at home as it requires expensive equipment, so I'll just describe it briefly. I followed instructions obtained from a research paper by Kim Penn and William V. Slaton, "Measuring Model Rocket Engine Thrust Curves," which was published in *The Physics Teacher.*

This method uses an engine bracket that you can build or purchase, a force probe, a computer interface box, Logger Pro software, and a laptop. I used a Vernier force probe linked to a Vernier Labpro interface box with the Logger Pro software on a laptop. This method involves quite a bit of extensive hardware, so I don't describe how to construct the engine bracket. You can see my setup in Figure 8-11. Although on a smaller scale, you can see the bracket is similar to the holder you built for the digital-balance method.

I first tested an A3-4T engine. My graph in Logger Pro shows a quick force spike followed by a constant stable force of about 2 Newtons. The software calculated the impulse at 2.32 Newtons-seconds, which is close to the published impulse for this engine. The National Association of Rocketry lists this engine at 2.22 Newtons-seconds of impulse.

I also wanted to test my larger engine so I conducted the experiment with a pair of B4-4 engines. You can see the resulting graph in Figure 8-12 and the video on the book's website.

Of the two methods to test the thrust of the rocket engine, using the force-probe method is probably preferable, though the digital-scale method gives good results, too, and is a useful alternative. Mass is released during the burning of the engine, and this can cause errors when using the digital-scale method, however.

FIGURE 8-11 Testing a model rocket engine using an engine bracket and force probe

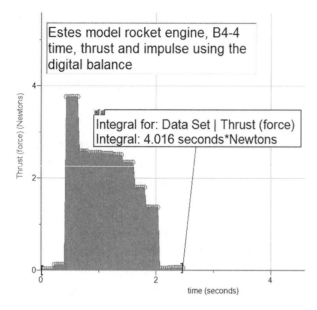

FIGURE 8-12 The force-impulse graph found with an Estes B4-4 model rocket engine

Summary

Momentum and impulse are physics concepts that help physicists analyze mass and speeds of many collisions. No matter the type of collision, momentum is always conserved: the momentum of the colliding objects before the collision is equal to the momentum of the objects after the collision. Also by changing the momentum of an object, you obtain an impulse. The impulse-momentum equation uses Newton's second law of motion. In the next chapter, you'll take much of what you've learned so far about motion and fly a few model rockets.

CHAPTER 9

Blast Off! The Physics of Rocketry

YOU'VE LEARNED AN INCREDIBLE AMOUNT of physics so far! You've investigated speed, distance, acceleration, energy, forces, momentum, projectiles, and free fall. Rockets make a good final chapter that brings together many of these physics concepts. In this chapter, you assemble a rocket launch ignition system and construct a rocket launch pad before putting together several rockets. Then using physics equations and video analysis, you'll figure out how high your rockets travel, how fast they move, and their rate of acceleration. Rocketry is an incredibly complicated science that involves a generous plateful of mathematics, physics, chemistry, and engineering. But with the tools you have, you can make good approximations for speed, altitude, and acceleration. We've got a lot of ground to cover, so let's get started right away with the first project!

Project: Build a Rocket Launch Ignition System

In Chapter 8, you learned about momentum and engine thrust. You built a rocket engine holder and used a digital scale and a video camera to get the impulse values for your rocket engines. Your first project for this chapter is to build your own model rocket launch ignition system.

Rocket Safety

The National Association of Rocketry (NAR) publishes guidelines for launching rockets safely. Among the rules are guidelines about the launching system:

> I will launch my rockets with an electrical launch system and electrical motor igniters. My launch system will have a safety interlock in series with the launch switch, and will use a launch switch that returns to the "off" position when released. (http://www.nar.org/NARmrsc.html)

The ignition system first must be "turned on" and then another switch must be depressed to launch the rocket. This secondary switch must also return to the off position after the rocket is launched. The NRA states that rockets with small engines (sizes A–D) must be launched at a distance of 15 feet away.

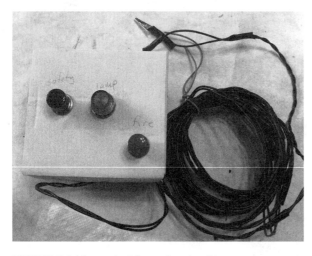

FIGURE 9-1 My rocket launcher ignition system

Although assembling a model rocket launcher ignition system from parts might cost more than the $10 to $15 for a commercial product, building your own is more fun. And you might already have some of the materials you need at home; if not, you can purchase items for this project at hardware, hobby, and/or electronics parts stores or through online sources. You can see my finished rocket launcher ignition system in Figure 9-1.

Things You'll Need

Parts

- Small box for an enclosure. This could be a shoebox, small cardboard box, or other type of enclosure. I used a small box that was about 4" × 4" × 1½" deep.
- A package of two SPST momentary push-button switches. I found mine at Radio Shack (part #275-0609). You'll use both, one as your "safety" switch and the other as your "fire" button.
- T-3 ¼" miniature bayonet lamp base. I got mine at Radio Shack (part #272-325).
- Two 6.3 V (Volt), 150 mA (milliampere) T-3 ¼ bulbs.
- Two mini (1¼") alligator clips (they often come in packages of 12).

- ¾" or equivalent electrical tape.
- One fresh package of 4 AA batteries.
- 4-AA battery holder with leads.
- Stranded, 22-gauge electrical hookup wire, at least 40 feet.

Tools

- Drill with ¼", ½", and 2 $\frac{1}{32}$ " drill bits
- Wire stripper and wire cutters

NOTE

Often these are part of the same tool, but you need both—the stripper to remove the wire insulation and the cutter to cut the wire.

- Hot glue and hot glue gun
- Solder gun and thin-wire electrical solder
- Ruler
- Scissors
- Needle-nose pliers
- Masking tape
- An extra model engine igniter for testing
- Multimeter for testing
- Eye protection

BE CAREFUL!

Follow safety guidelines when using the drill and other tools. The hot glue and solder guns can get extremely hot and cause burns. Work under the supervision and with the assistance of an adult.

1. Cut two long lengths of wire, each approximately 20 feet long. Two of the ends of these wires will go to the engine igniter.
2. Braid the wires together, as shown next. Don't braid too tightly as you need at least 15 feet of wire for a safe launch. Leave the last four

inches from the end unbraided. Wrap the wire with some electrical tape at 6-inch to 1-foot intervals.

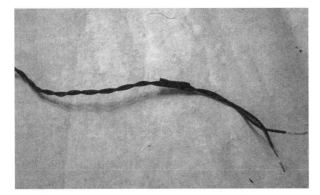

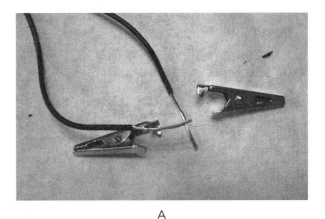

A

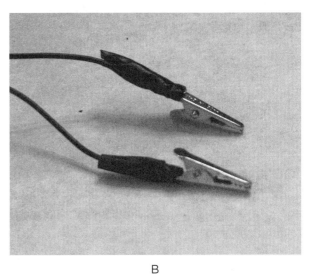

B

3. Use the wire strippers to strip the insulation off about ½" of wire from each end of both wires.

4. Open the alligator clip package and remove two clips. Run the stripped wire through the hole of the alligator clip, one wire for each clip. Crimp using needle-nose pliers, wrapping the extra wire around the clip. Then solder the wire to strengthen the attachment and cover it with electrical tape. You can see this process illustrated in Figure 9-2. The two alligator clips hold the two ends of the igniter in the rocket engine.

FIGURE 9-2 Running the stripped wire through the hole of the alligator clip (A) and the finished clips (B), soldered and covered with electrical tape

 BE CAREFUL!

You're using the solder and solder gun to strengthen electrical attachments (both here and later in the project). Work carefully with this tool as a solder gun gets extremely hot. And always work under adult supervision.

5. Take the box for your enclosure. Here you can see a diagram of the holes you'll drill in the front (or top) of your box. Drill two holes with the ½" drill bit (these are for the safety and fire switches) and drill one hole with the $2\frac{1}{32}$" drill bit for the lamp.

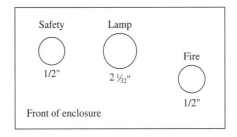

6. Drill two holes in the bottom of the box with the ¼" bit. These smaller holes are for the two long ignition wires. Figure 9-3 is a picture of the box with all the holes drilled.

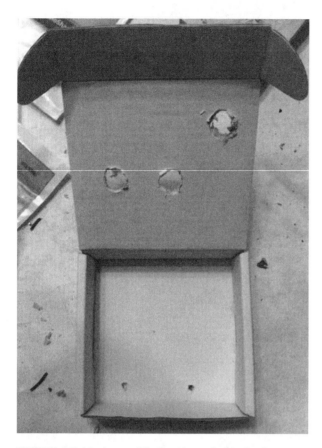

FIGURE 9-3 My box with the three holes in the top and two smaller holes in the bottom of the enclosure

7. Label the holes at the top of your enclosure, as shown here. The first hole should be labeled **Safety**, the second **Lamp**, and the third **Fire**.

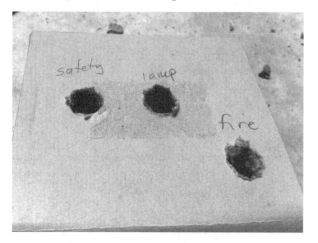

8. Open the momentary push-button switch package and remove both switches along with the hardware. Open the lamp base package, too.

9. One of the switches will be the Safety switch. Push it through the top of the enclosure in the hole labeled "Safety," and reattach the washer and nut from below and tighten, as shown in Figure 9-4.

10. Insert the lamp base and the other switch (the fire switch) in their holes. Remove one of the lamps and insert it into the lamp base and put on one of the covers (either green or red, your choice) (Figure 9-4).

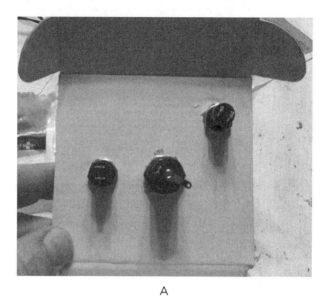

A

B

FIGURE 9-4 A bottom (A) and top (B) view of the enclosure with the switches and lamp in place

11. Open the battery holder. The battery holder has a red wire (positive) and a black wire (negative) already attached. These wires are already stripped, but you might find it helpful to strip a little more from each end.

12. Cut three pieces of wire, about five or six inches each, and strip about ½" off the ends of all three pieces.

13. Now it's time to wire your components. Study the diagram in Figure 9-5 and use it to attach your wires. The diagram shows the components as seen from below. The safety, lamp, and fire components each have two terminals. The inner circles represent the terminals. You'll use the solder and solder gun to strengthen the electrical attachments. Remember the solder gun gets extremely hot. Make sure an adult is present to supervise. Here are the steps for wiring the switches and lamp base terminals:

a. Attach the red wire of the battery holder to one of the terminals of the fire switch. On this same terminal attach one end of one of your stripped 6-inch wires. Securely attach these wires to this one terminal of the fire switch. Solder and cover securely with electrical tape.

b. Take the other stripped end of this same 6-inch wire and attach it to one terminal of the lamp base. Solder and cover with electrical tape.

c. Take another of your 6-inch wires and attach one end of it to the other terminal of your fire switch. Attach the other end of this 6-inch wire to the other terminal of your lamp base.

d. Attach one end of the last 6-inch wire to the same lamp base terminal and the other end to one of the safety switch terminals. Make sure to solder and cover these connections securely with electrical tape.

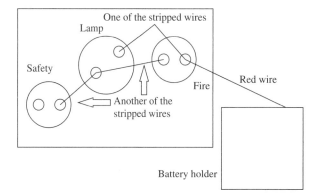

FIGURE 9-5 Diagram for wiring the switches and lamp base terminals

14. Take the two ends of the long wires (the ends without the alligator clips) and feed them through the two small holes you made in the bottom of the enclosure. Figure 9-6 shows the diagram with the long wires added.

15. Take the black wire from the battery holder and attach it to the stripped end of one of the long wires.

16. Take the end of the other long wire and attach it to the other terminal of the safety switch. Again, solder and add electrical tape to the connections.

Before you finish, let's run a few tests. Place a fresh set of four AA alkaline batteries correctly into the battery holder. Take the two alligator clips and

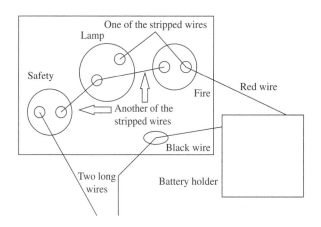

FIGURE 9-6 Diagram for wiring the two long wires into the enclosure

hook them together. Press the safety switch and the test lamp should light up. Unhook your alligator clips from each other.

BE CAREFUL!

Make sure you have your eye protection on and complete this test with an adult present. Work in a safe area without anything flammable around.

If you have an extra model rocket igniter, attach it to the alligator clips (one igniter wire on each alligator clip). When you attach the igniter to the alligator clips and press the safety switch, the lamp should light up, as shown in Figure 9-7. You are making a complete circuit but are not feeding the igniter enough current to ignite it. Now hold down the fire push-button switch while pressing the safety switch. The igniter should ignite as you are sending enough current through the igniter to activate it. The lamp should now go out as you have broken the pathway (circuit).

You can perform several additional tests with a multimeter, as shown in Figure 9-8. First, attach

FIGURE 9-7 Testing the launch system with an igniter

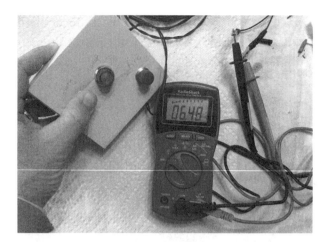

FIGURE 9-8 Testing the voltage of your launch system. Mine tested at 6.48 Volts.

your two multimeter leads to the alligator clips of your launch ignition system. It doesn't matter which multimeter lead goes to which alligator clip. Set up your multimeter to measure DC voltages. Press the safety switch and the multimeter should read around 6 to 6.5 Volts (given four new AA batteries). Each AA battery is rated at 1.5 Volts so four AA batteries should be equal to around 6 Volts.

You can also test the current (charge flow). *Current* is the flow of charge, in this case, the motion of electrons moving in the wire. Current is measured in *amperes* (or *amps* for short). Keep the multimeter leads on the alligator clips but make sure to connect the leads correctly into the multimeter unit to measure current. On my multimeter, I had to move the red lead to the other port to measure current (amps) and set the multimeter to measure DC current. When you have the multimeter set up correctly, press the safety and the fire buttons at the same time. The multimeter is now measuring the current available to light the igniter. My testing gave my launch system a current of slightly over 4 Amperes (Amps).

If your tests did not work, check your wiring and make sure your connections are soldered well and there are no *shorts* (wires touching that shouldn't) in the system. Make sure to use new alkaline batteries for the best results. If the lamp doesn't light when pressing the safety button, then you don't have a complete pathway; there must be a short or a wire may not be making a good connection. Check your wiring against the wiring diagrams in Figures 9-5 and 9-6.

My cardboard box enclosure was not deep enough to close it with the battery pack inside, so I taped my batteries and battery holder to the back side of my box, as shown here.

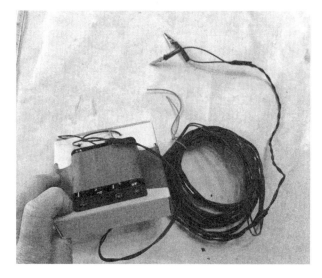

You can use tape or hot glue. If you used a larger box, then everything can fit inside. You'll use it later in the chapter to launch your rockets. Notice that I coiled my launch wires up with another wire, or you could use a twist-tie to hold the wires when the system is not in use.

Project: Build Your Model Rocket Launch Pad

The next project is to build a launch pad for our model rockets. Commercial products are available (Portapad, for example), but you can build your own fairly inexpensively. I found several do-it-yourself launch pad options online. One set of plans created by Stefan Jones (http://makeprojects .com/Project/Portable-Model-Rocket-Launch-

Pad/1039/) served as a model for mine, although I did make modifications to the original plan.

The National Rocketry Association recommends a launch pad with a guide rail that helps keep the rocket vertical or nearly vertical as it starts to launch. The Association also recommends using a blast deflector that redirects the rocket engine's exhaust so it doesn't hit the ground directly. Finally, the guide rod must be removed when not in use or remain at a height above eye level to protect against eye injury or be capped in some way.

Things You'll Need

Parts

- Three 2'-long straight tubes of PVC pipe, each ¾" in diameter (You can use 1' or 2' sections, whichever is more readily available.)
- Three ¾" PVC "T" joints
- Three 45-degree ¾" PVC joints
- One 3'-long steel 1/8" rod (The label on the rod I purchased described it as a "1/8, Music Wire.")
- Epoxy
- Square piece of wood (I used a 2 ½" square that was ¾" thick; however, the measurements do not have to be exact.)
- Rubber bands
- Aluminum or stainless-steel pie pan—but do not use a disposable aluminum pie pan
- Two metal clothes hangers
- Paper plate
- Several plastic spoons to help mix the epoxy
- Round wood ball

Tools

- Drill and 9/64" drill bit
- Hot glue gun and glue sticks
- Pliers or needle-nose pliers to cut and bend the clothes hangers

BE CAREFUL!

Take care when using the drill and hot glue gun. Use the epoxy in a well-ventilated space. Always work with an adult to supervise.

1. Take the three 45-degree PVC joints and place on a paper plate. Make sure they are flat with the joints splayed out as shown here. Attach them as one unit with several rubber bands.

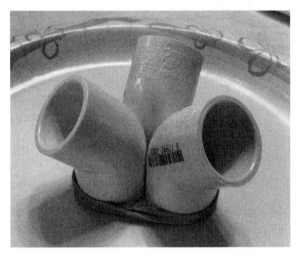

2. Mix up some of the epoxy and spread this between the joints. Keep the rubber bands on to hold the joints in place while the epoxy is drying.

3. When the epoxy has dried, remove the rubber bands and flip the joints over and epoxy around and in between the joints on the side, too. Let them dry. Don't use up all the epoxy as you'll need some later.

4. Push the T joints on the PVC pipe so they fit snugly. These will be the legs of the launch pad.

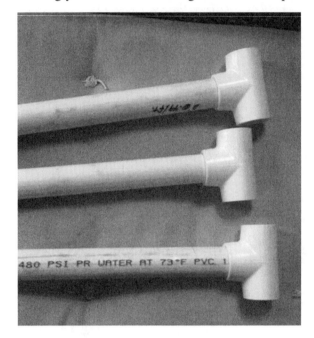

5. While the joints are drying, you'll work on the deflector shield. Take the pie pan and set it down interior side up. Place the square of wood in its center. Hot glue this square piece of wood to the pie pan at its center.

6. After the hot glue has dried, place a small dot at the center of the piece of wood and use a drill with a ⁹⁄₆₄" drill bit to drill through the wooden square and pie pan.

7. When the epoxy on the three joints has dried, take the same drill bit and drill through the epoxy to make a hole through the center of the epoxy. When finished with the drilling, run the steel 1/8" guide rod through the epoxy hole. The rod should fit snugly.

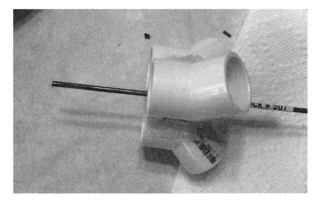

8. Now line up the drilled holes in the epoxy of the joints and in the square of wood and pie pan with the guide rod. Epoxy the circular edges of the joints to the wood square. Keep the guide rod in place while the epoxy is drying to keep the drilled holes lined up.

9. Take one of the clothes hangers and make two cuts, as shown in the picture here. Bend the curved area of your cut piece with your pliers to make a more rounded bend. You need three of these in total as the stakes for the PVC legs to add stability to the rocket launcher during our launches.

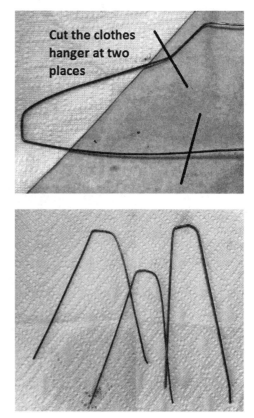

Cut the clothes hanger at two places

10. Use a wood ball to cap the guide rod when not in use. Using the ⁹⁄₆₄" drill bit, drill partway into the wood ball. The guide rod should fit into this hole snugly. This caps the rod to protect you from accidental eye injury when not in use.

11. Let's put everything together. Attach the three legs to the three joint holes and configure them to make a tripod. Make sure the flat surface of the pie pan is on top. Put the guide rod through the holes and cap the top of the rod with the wood ball. The rod should go through the holes and you can adjust its height by pushing it further down. Figure 9-9 shows the completed launch pad.

You can disassemble the launch pad for easy transport. You've got the guide rod, pie pan apparatus, the three legs, and the three stakes. I transported the disassembled launch pad to my community's RC airplane and model rocket park and then reassembled the pieces at the park. Keep the end cap of the guide rod capped when not in use.

Assemble Your Rocket Kits

You have your launcher ignition system and your launch pad. Now you need a rocket. You have several options. You can build the entire rocket from scratch, buy a rocket kit with all the parts but it has to be assembled, or purchase a ready-to-fly rocket. I went to a local hobby store and found several kits, buying a larger rocket and a smaller rocket.

The larger rocket I built is called the "Baby Bertha," and the smaller rocket is called the "Yankee" (Figure 9-10). Each kit was about $9.

A

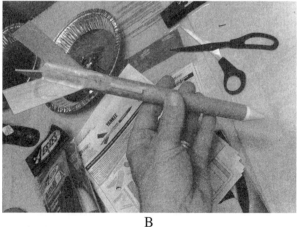

B

FIGURE 9-9 The completed launch pad with rocket

FIGURE 9-10 (A) The completed "Baby Bertha" model rocket kit pre-paint and decals and (B) the "Yankee" model rocket being assembled

All the parts to build each rocket were included. It took several hours to build each, plus you'll need glue and a hobby knife.

The packaging has information about engine requirements for each rocket. Most rockets are designed to hold different engines. However, the company recommends a certain engine for the first flight. For both of my rockets, the Baby Bertha and Yankee, they recommend an A8-3 engine for their first flight. The packaging also contains information about projected altitude and mass for the rockets.

Experiment
Physics with Rockets

Launching rockets can be great fun, but you can use the ignition system and launch pad to launch the rockets and then apply the physics you've learned in earlier chapters to come up with the rockets' acceleration, speed, and height. You won't be able to get exact numbers because of the complexity of the equations and variables, but you'll get an idea of how you can use video analysis coupled with some of the equations you've learned to get approximate figures.

Before taking the rockets to the park, buy engines and extra igniters and the recovery wadding. The wadding goes into the rocket body to protect the parachute from some of the hot gases coming from the engine. For my experiments, I purchased three engines: an A8-3, B4-4, and C6-5.

Find the Acceleration with Video and Tracker

On the day of my rocket tests, I fired off both the Baby Bertha and the Yankee rockets. In the Baby Bertha, I first used an A8-3 engine, followed by a B4-4, and then a C6-5. I launched the Yankee rocket with the A8-3 engine.

When filming, use a tripod and set it several meters away from the launch pad. The goal is to get video of

BE CAREFUL!

Follow area laws when going out to shoot your model rockets. The National Association of Rocketry has a list of helpful safety rules and recommendations on their website: http://www.nar.org/NARmrsc.html. Finding a safe, large open space may be a challenge. Consider a local park or model airplane facility. Remember to bring eye protection and extra supplies such as masking tape, glue, and AA batteries. Have an adult present to supervise and give you a hand. It is helpful to have an extra set of hands and eyes to help run the equipment, fire off the rockets, and then find them!

the rocket accelerating upward from the launch pad. Back off enough to get at least five or six frames, but you also want to be close enough to see the part on the rocket that you'll "frame out" for Tracker.

Also make sure you have a meter stick in the field of view or something to use as the calibration stick for the video. One way to do this is to measure the height of your rocket launch from the ground to the deflection pad, and use this measurement as the calibration stick in your analysis. To aid your analysis, place some blue masking tape around the rocket under the nose.

When you return home, upload the separate videos into your computer and combine them into one video for analysis with the Tracker software. I included my video on the website. Feel free to use it, or, better yet, create your own video for your rocket's launch!

Acceleration is rate of change of speed or velocity and represents a change of speed or velocity with respect to time. Frame out your video. For mine, I tracked frames 255–261 of the video with a frame step size of 1. I used the launch pad as my calibration stick (50 cm) and framed a point on the

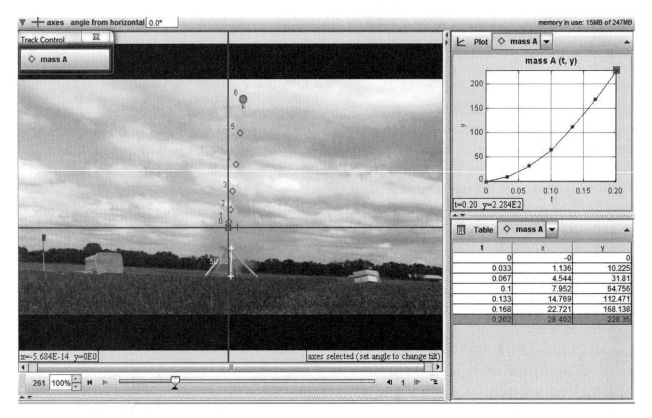

FIGURE 9-11 Framing out the video of launching the Baby Bertha rocket with an A8-3 engine

rocket I could easily see in the video (the blue strip under the nose cone). Figure 9-11 is a screenshot from framing out my video in Tracker.

You can see my rocket is accelerating from the parabola on the distance-time graph shown in Figure 9-12. To determine the acceleration, create a speed-time graph (t, v_y). This graph should show a diagonal line. Now select **Data Tool (Analyze…)** and **Fit**. A linear fit gives you the slope of the graph. This slope is a measure of the acceleration. You can find this acceleration on the Data Tool screen and circled in Figure 9-12. For my rocket, I obtained an acceleration of 9576.57 centimeters per second per second (cm/sec/sec), which converts to a rounded 96 meters/sec/sec. I replicated the Tracker analysis for the smaller Yankee rocket with

the A8-3 engine and obtained an acceleration of 209 meters/sec/sec. Depending on the rocket and engines used in your launches, your acceleration rate may be close to or somewhat different than mine.

Using Newton to Find the Acceleration

For a rocket, three general forces act together to create the acceleration: the thrust, the air resistance (often called *drag*), and the rocket's weight ($W = mg$). The thrust is upward and the drag and the rocket's weight are downward forces. In the case of the rocket's initial motion, the thrust is a much greater force than the other two forces and this creates the upward acceleration on the rocket.

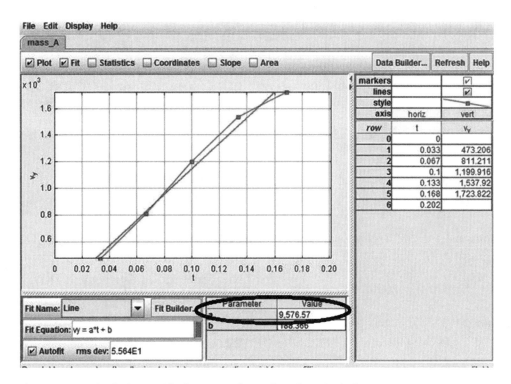

FIGURE 9-12 The Data Tool window with the rate of acceleration circled

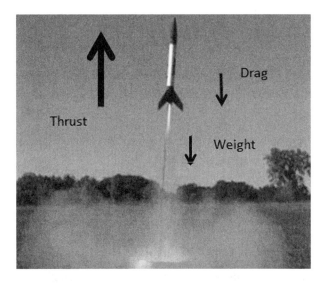

Newton's second law, which you learned about in Chapter 6, helps determine the rocket's acceleration. Newton's second law states that the net force on an object is equal to the object's mass (*m*) multiplied by its acceleration. In symbolic form, you would write:

$$F_{net} = m(a)$$

For your rocket, the thrust is upward, the drag (air resistance) is downward, and the weight is downward. So you can write Newton's second law as follows:

Thrust – Drag – Weight = mass of the rocket × rocket's acceleration

Because you can't really determine the drag easily, you'll omit this term. Let's look at the smaller rocket, the Yankee. I flew the Yankee with the A8-3 engine. If I look at the manufacturer's thrust-time graph for this engine, I see that the thrust for this engine builds from 0 Newtons to around 10 Newtons (http://www.321rockets.com/a8-3-estes-rocket-engines-1598.html). This thrust buildup proceeds fairly linearly based on the manufacturer's thrust-time graph. So you can take half of the 10 Newtons for the average thrust (which is 5 Newtons) through this time segment.

Using the digital balance from Chapter 8, you can obtain the mass of the rocket. You can find the engine's mass from the engine specifications. The weight of the rocket is the *mass* (rocket + engine = 14 grams + 16.2 grams = 0.0302 kg) multiplied by little *g* (9.8). Rewrite the equation and add some numbers:

Thrust – Weight = mass of the rocket × (rocket's acceleration)

$$5 \text{ (Newtons)} - 0.0302 \text{ (kg)} \times 9.8 \text{ (m/s/s)} = 0.0302 \text{ (kg)} \times a$$

$$5 \text{ (Newtons)} - 0.29596 \text{ (Newtons)} = 0.0302 \text{ (kg)} \times a$$

$$4.70404 \text{ (Newtons)} = 0.0302 \text{ (kg)} \times a$$

$$\text{Acceleration} = 155.76 \text{ meters per second per second}$$

Tracker software gave my rocket a little larger acceleration at 209 m/s/s, but using Newton's second law I'm in the ballpark. Both methods—using video and Tracker, or using Newton's second law—rely on some assumptions: you don't really have a constant acceleration and you're leaving out air resistance. Another important point is the rocket loses mass as the engine burns its fuel. In this equation, you assume that the mass stays constant. This equation is helpful, but it can only give approximate answers!

Let's Calculate Speeds

One of the things to determine about a moving object is the speed it has at some time or point along its path of motion. For the rocket, you can find several speeds. For example, how fast is the rocket going by the time it has left the field of view in your video for Tracker? This speed can easily be determined by knowing the time and the acceleration. Let's take a look at Baby Bertha with the A8-3 engine.

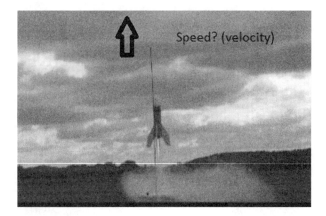

Going back to the Tracker analysis, I analyzed frames 255 to 261, for a total of 6 steps if I start with frame 255 as data point 0. The time between each frame is 1/30 of a second or a total time of 6/30 of a second or 0.2 seconds.

By taking the acceleration and time, you can determine the speed of the rocket. In Chapter 2, you learned, this equation:

change of speed of an object = (its constant acceleration) × (time)

In symbolic form, you can write it like this:

$$v_{final} - v_{initial} = a(t)$$

That is, a speed at some point in time (*final speed*) minus *initial speed* is equal to the product of *acceleration* and *time*.

For the rocket, you want to calculate its final speed. The initial speed of the rocket when it starts is 0 as the rocket is at rest.

$$v_{final} = at = 96 \text{ m/s/s} \times (0.2s) = 19.2 \text{ m/s of speed at the top frame of the Tracker analysis}$$

This speed—19.2 m/s—is pretty fast, isn't it? It's equal to around 43 miles per hour! The rocket engine propelled the rocket from a stationary state to 43 miles per hour in 0.2 seconds. What speed did you get for your rocket? As with other analysis, this method does have some potential problems. This method assumes you have a constant acceleration for this time duration (0.2 seconds), which isn't exactly true. This equation does, however, give you an estimated speed.

The Maximum Speed of the Baby Bertha and Its A8-3 Engine

But the speed of 19.2 m/s is not the maximum speed for the Baby Bertha rocket. The rocket is still potentially being accelerated past the final frame in Tracker. The rocket gains speed because of thrust, so it should theoretically gain speed up to the point where the rocket's engine gives out. In reality, though, as the rocket gains speed, air resistance (drag) works against the thrust and decreases the rocket's acceleration. But let's ignore the air resistance and see if you can find your rocket's theoretical maximum speed.

One method is to use the impulse-momentum equation, or theorem, that you learned about in the last chapter: $Ft = m(v_{final}) - m(v_{initial})$. For this chapter, I repeated an impulse test using the A8-3 engine, replicating the procedure developed in the previous chapter using the engine bracket and digital scale. Try it with your rocket. I ignited my A8-3 engine and filmed video of the digital scale readings and then converted my data into a thrust-time graph, as shown here.

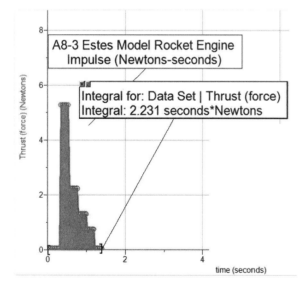

When you look at the graph for my A8-3 engine, you see that I obtained an impulse of 2.2 Newtons-seconds using the digital scale method (the engine manufacturer reports an impulse of 2.5). Using my impulse, you can write:

$$Impulse = (mv_{final} - mv_{initial})$$

$$2.2 = mass\ of\ rocket \times (final\ or\ maximum\ speed) - mass\ of\ rocket \times (initial\ speed)$$

The rocket starts from rest so its initial speed equals 0. I determined the mass of Baby Bertha without an engine to be 46 grams. The A8-3 engine has a reported total mass of 16.2 grams, containing a propellant mass of 3.12 grams, so the casing has a mass of 13.08 grams (16.2 – 3.12 grams = 13.08 grams). This means the Baby Bertha starts out with a mass of 62.2 grams (46 + 16.2 grams), and at its maximum speed, it should have a mass of 59.08 grams (46 + 13.08 grams) as it uses its propellant. If you take an average of these two values, you get a mass of 60.64 grams. You'll use this value for the mass of the rocket (0.06064 kilograms):

$$2.2 = 0.06064 \times (maximum\ speed)$$

So you can calculate the maximum speed, which equals 36.3 meters per second, or around 81 miles per hour.

Is this a believable speed? I would guess that it's a bit high because you ignored any air resistance on the rocket. With air resistance, maximum speed will be smaller, as this force is pushing against the rocket. This change of speed is based on the impulse, and the digital scale method of finding the impulse is uncertain, too. Finally, I averaged the masses. This method does give an estimated speed, however. What was your estimate for your rocket's maximum speed?

Determine the Rocket's Altitude

The rocket achieves its altitude in two separate ways that you can combine to determine its altitude or total height. When the rocket burns its fuel and accelerates, its thrust provides a speed and height. This is similar to a car accelerating up to the speed limit after the stoplight turns green. Once at the speed limit, the driver can take his or her foot off the gas pedal and coast to a stop. The car and driver continue to move forward during coasting.

The rocket's acceleration has given it a height and speed. At the moment the engine gives out, the rocket has an upward speed and now begins to coast farther upward. You need to calculate this additional height and add it to the height achieved by the acceleration. The rocket during this coasting phase is a free fall object traveling upward—it is just like the tennis ball after it is shot by the tennis ball cannon you constructed in Chapter 3.

Let's first calculate the height achieved by my rocket's engine. If I know the acceleration for the Baby Bertha with the A8-3 engine and I know its maximum speed, I can use these two bits of information to tell me both the time this occurs along the flight path and how high above the launch pad this occurs. Let's use this equation again:

change of speed of an object = (its constant acceleration) × (time)

$$v_{final} - v_{initial} = a(t)$$

Earlier, I calculated the maximum speed (v_{final}) at the point of engine burnout to be 36.3 m/s, my initial speed is 0 for the rocket, and my acceleration as found by Tracker was 96 m/s/s. Then I can solve for time:

$$36.3 \text{ m/s} = 0 \text{ m/s} + 96 \text{ m/s/s}(t)$$

The time to get the rocket up to this maximum speed is 0.38 seconds. This seems reasonable based on some of the assumptions stated earlier.

If you know the time, you can predict how high the rocket will be above the launch pad at the point when the rocket starts to coast. The following equation is a *kinematics equation* designed to find how far (distance) an accelerating object travels:

Distance = ½(acceleration)(time)² + $v_{initial}$(time)

Distance = ½(96 m/s/s)(0.38s)² + 0 (as $v_{initial}$ = 0)

If you solve for distance, it appears that my rocket is around 7 meters above the launch pad at this point. This number represents the height achieved

through acceleration from the engine. Figure 9-13 should clarify what you know about the rocket's motion so far.

The Coasting Phase of Your Rocket

When the rocket is coasting upward, the acceleration on the rocket is now 9.8 m/s/s acting downward. This represents the full effect of gravity on the rocket. For this equation, you call this a negative number. The rocket has a starting speed of 36.3 m/s, and it will eventually be stopped by gravity; this occurs at the apex or the top part of its path. Slowing the rocket takes time. You can find this time by reusing the equation:

$$v_{final} - v_{initial} = a(t) \text{ or}$$

$$v_{final} = v_{initial} + a(t)$$

$$0 \text{ m/s} = 36.3 \text{ m/s} - 9.8 \text{ m/s/s (time)}$$

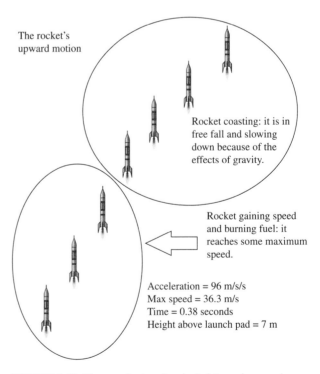

The rocket's upward motion

Rocket coasting: it is in free fall and slowing down because of the effects of gravity.

Rocket gaining speed and burning fuel: it reaches some maximum speed.

Acceleration = 96 m/s/s
Max speed = 36.3 m/s
Time = 0.38 seconds
Height above launch pad = 7 m

FIGURE 9-13 The rocket gains height and speed through burning its fuel. It gains 7 meters of height because of acceleration from the thrust produced by its engine.

So time = 3.7 seconds (the rocket coasts for 3.7 seconds).

How much distance will it take to slow the rocket down? Reuse the distance equation, but now you need a term for the initial speed:

Distance = ½(*acceleration*)(*time*)² + $v_{initial}$(*t*)

Distance = ½(–9.8 m/s/s)(3.7s)² + 36.3 m/s(3.7s)

Distance = 134.31 m – 67.081 m = 67.2 meters

You can now add both distances to get the total altitude achieved by the rocket. It moved 7 meters during the engine-burning phase and an additional 67.2 meters during the coasting phase. This gives you a total of 74.2 meters for the altitude of the rocket above the launch pad. This is equivalent to 243 feet. Is this a believable number? In Figure 9-14, you can see the coast motion has been added to Figure 9-13.

The Baby Bertha packaging mentions that the rocket might achieve an altitude of

175 meters or 575 feet maximum. This maximum altitude is projected with a C engine, however. From conducting more online research, I got estimations of 100–200 feet with an A engine. The mathematical calculation of 243 feet is a believable number. Remember, too, my calculations ignore air resistance, so 243 is probably somewhat high.

Were you able to progress through all of the calculations for one of your rockets? The goal is to obtain a rocket's acceleration, calculate several of its speeds, and then analyze its altitude, all through some basic physics equations and video analysis. Use Tracker to obtain an acceleration and then use this acceleration and the engine impulse to determine rocket speeds. Coupled with times of flight, you proceed to find several heights that you add to find the total altitude. Whew, lots of numbers and calculations!

Famous Scientists

Robert Goddard was one of the first American scientists to explore the physics and engineering of rocketry in the early and mid-1900s. Along with the Russian scientist Konstantin Tsiolkoysky and the German Herman Oberth, Goddard is considered one of the pioneers of modern rocketry.

As a boy growing up in Massachusetts, Robert Goddard was interested in science and technology. He read H. G. Wells' *The War of the Worlds* as a teenager and dreamed of space travel through the atmosphere and to the planets and moon. His father reinforced Goddard's love of science with gifts of a microscope, telescope, and science books and magazines. Robert Goddard built kites and balloons as a boy. In school, he read Isaac Newton and became especially focused on Newton's third law of motion—action-reaction— and how this scientific principle could be used to propel rockets upward.

Robert Goddard studied physics at the Worcester, Massachusetts, Polytechnic Institute, and received a degree in physics in 1908. He completed his graduate studies in physics in 1911. He studied ways to stabilize airplanes and rockets through the

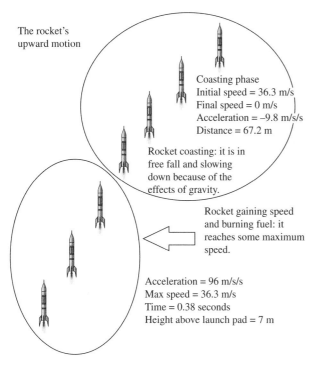

The rocket's upward motion

Coasting phase
Initial speed = 36.3 m/s
Final speed = 0 m/s
Acceleration = –9.8 m/s/s
Distance = 67.2 m

Rocket coasting: it is in free fall and slowing down because of the effects of gravity.

Rocket gaining speed and burning fuel: it reaches some maximum speed.

Acceleration = 96 m/s/s
Max speed = 36.3 m/s
Time = 0.38 seconds
Height above launch pad = 7 m

FIGURE 9-14 Coasting motion added

use of gyroscopes and researched different rocket fuels and how to construct multistage rockets. He realized, too, that rocket fuels did not have to be solid but could also be liquid. Here you can see a photo of Goddard with one of his liquid-fueled rockets.

Many scientists scoffed at Goddard's ideas and believed that studying rockets was a waste of time with very little practical use. Goddard persevered, however, by building and launching more powerful rockets and studying how to make them more efficient and to fly higher. He developed many of the calculations used today to determine the motion of a rocket, using Newton's calculus and physics equations. When Goddard wrote an article about sending a rocket to the moon and into space, many people laughed. Goddard would not live to see his ideas come to reality, as he died in 1945. Today, he is seen as a forward-thinking scientist whose ideas made modern rocketry and spaceflight possible.

INTERESTING FACT

You have probably seen Einstein's equation $E = mc^2$. It's one of the most famous equations in the world. It tells us that mass and energy are interchangeable. Under the right conditions, mass (or matter) can be converted to energy and vice versa. This conversion happens naturally to keep stars shining, and as humans, we've used it to build atomic bombs and nuclear power plants. Many individuals believe that the key to long-duration space flight is Einstein's famous equation. Both nuclear fusion and matter-antimatter annihilation are possible rocket propulsion systems that will give our future rockets large enough accelerations and speeds to journey to far-away planets and star systems.

Summary

In this chapter you built an ignition system and a launch pad, and then you built some rocket kits. Launching model rockets is fun! Even more impressive is how you can use physics equations to estimate acceleration, speed, and height for your rocket. Isn't it amazing that physics equations can yield sensible answers to questions such as, "How high did my rocket go?" and "How fast did it go?"

I hope you enjoyed not only this chapter, but also the book as a whole! The concepts presented in this book only touch the surface of the science of physics. Continue to study math and science and remain curious about how the world and universe work!

APPENDIX

Resources

YOU CAN FIND A LOT OF HELPFUL GENERAL physics reference material online. Of course, your first stop online should be the book's website at *http://www.mhprofessional.com/fun_physics*. The Physics Classroom (*http://www.physicsclassroom .com/*) is a great resource. Another good one is HyperPhysics (*http://hyperphysics.phy-astr.gsu. edu/hbase/hframe.html*). Depending on the concept you want to learn more about, YouTube (*http:// www.youtube.com/*) has some instructive videos. If you enjoy building projects, see Instructables (*http://www.instructables.com/*) for many ideas. Thames and Kosmos also has science and physics projects and kits (*http://www.thamesandkosmos .com/*). I often use my iPhone and a small video camera like Flip Video Camera to take videos of my projects. The computer, of course, is essential in my classroom. We use software to make graphs and analyze our physics videos. Here is additional information about some essential software tools:

- Excel is a spreadsheet package that comes with Microsoft Office. Go to *http://office.microsoft .com/en-us/buy/?CTT=97* for more information. A student version is available at special pricing, but even so, it is pretty expensive. Your school may provide you with access to a copy, however.

- Calc is a free open source spreadsheet program from OpenOffice, which you can find at *http:// www.openoffice.org/*. It operates similarly to Excel.

- An alternative to the graphing functions found in Excel is a free program called Graph, which you can find at *http://www.padowan.dk/graph/*. Using this program, you can graph points and determine different equations and functions for a series of data points. This is an excellent free program to use in place of Excel.

- Tracker Video Analysis is a free video analysis software package, created and copyrighted by Douglas Brown. You can find it at *http://www .cabrillo.edu/~dbrown/tracker/*. When installing Tracker, you may find it helpful to install Xuggle, too, which can be downloaded with Tracker. Xuggle allows you to analyze more video formats.

- Finally, Logger Pro is a science software package used by many high schools and universities. Produced by Vernier Software and Technology, it can analyze video and control different science probes and sensors. It also allows you to create graphs and mathematically manipulate data. You can download a 30-day demo version from *http://www.vernier.com/ products/software/lp/*.

Several of the projects call for soldering. Here are two websites that provide tutorials on soldering:

http://w6rec.com/duane/bmw/solder/

http://www.kingbass.com/soldering101.html

You can also search on YouTube for soldering tutorials.

INTERESTING FACT

CONVERTING UNITS OF MEASURE

A unit is a way to measure a concept. For example, in the United States, we use miles per hour (mph) as the common unit to measure the speeds of cars. Many other countries use kilometers per hour (kph). In physics classrooms, speeds are often measured in meters per second or centimeters per second. To move from one unit to another, you convert the unit from one measure to another, for example, 1 mile = 1.6 kilometers.

Tracker Video Analysis, which you'll use throughout the book, defaults to units of centimeters and centimeters per second. You will have to perform conversions to change your speed to other units (such as miles per hour). Here are a few good conversions to remember:

- 3600 seconds per hour
- 100 centimeters per 1 meter
- 1600 meters per mile

Cell phone apps are available to help you convert from one unit to another. OnlineConversion.com is a website that helps you make conversions from one unit to another:

http://www.onlineconversion.com/

Chapter 1—Cruise Control: Constant Speed

Physics is full of definitions and concepts. This page on the Physics Classroom website gives more information about kinematics:

http://www.physicsclassroom.com/Class/1DKin/U1L1a.cfm

Car Kits

The kit I used to build the constant-speed car came from the Kelvin Company at

http://www.kelvin.com

Thames and Kosmos make solar and electric car kits that you can use for this experiment as well. Those kits are on the Thames and Kosmos website at

http://thamesandkosmos.com/products/exploration/sm.html

and

http://thamesandkosmos.com/products/construction/ebv.html

The Electric Dragster, developed by Middlesex University in the United Kingdom and produced by 4M, uses CDs for wheels and can be assembled in less than one hour. You can find more about the Electric Dragster by going to Amazon.com and searching for **Electric Dragster**. Several different companies sell the car.

Arbor Scientific also stocks a prebuilt constant-speed car:

http://www.arborsci.com/constant-velocity-car

More Experiments

You can use constant-speed tubes to show motion. You can find more at

http://www.teachersource.com/Density/Viscosity/ViscosityofOilTubes.aspx

The Institute of Physics has a Practical Physics website with experiments for measuring the speed of an object through liquids:

http://www.practicalphysics.org/go/Experiment_235.html

http://www.practicalphysics.org/go/Experiment_234.html

Chapter 2–Stop and Go: Changing Speed

A standard measure of a car's acceleration is the time it takes to go from 0 to 60 miles per hour. Zero to 60 Times provides listings of cars and the time it takes them to accelerate:

http://www.zeroto60times.com/

Graphs and Parabolas

The Physics Classroom website has an excellent discussion of kinematics graphs, both *distance-time* (also called *position-time*) graphs and *speed-time* (also called *velocity-time*) graphs:

http://www.physicsclassroom.com/Class/1DKin/

The University of Colorado website provides an interactive simulation that explains distance, speed, and acceleration graphs:

http://phet.colorado.edu/en/simulation/moving-man

YouTube is a great source of videos that explain physics concepts. Here's one that explains the position versus time graph (be sure to look for more):

http://www.youtube.com/watch?v=4J-mUek-zGw

NOTE

If you're having trouble finding a part for the fan car, check out the "Car Kits" section for Chapter 1.

For more information about parabolas, see

- *http://www.algebra.com/algebra/homework/quadratic/*   Information about the parabola and quadratic equations
- *http://www.mathopenref.com/quadraticexplorer.html*   Change the A, B, and C constants to see how this affects the shape of the parabola

- *http://jwilson.coe.uga.edu/EMAT6680Su07/Singer/Assignment%202/Exploring%20Parabolas.html*   Information about the A, B, and C constants

Equations

If you would like to study the complete kinematic equations and their derivations, a number of websites and videos are available:

http://dev.physicslab.org/Document.aspx?doctype=3&filename=Kinematics_DerivationKinematicsEquations.xml

http://www.youtube.com/watch?v=gssrmSa7fLU

http://www.youtube.com/watch?v=nHmZEunBWgY&feature=related

http://www.youtube.com/watch?v=pD5jfwGe7eQ&feature=related

Chapter 3–Free Fall: What Goes Up, Must Come Down

I found the directions for the tennis ball cannon project on the Team Dandy website and have built several tennis ball cannons based on these instructions:

http://www.teamdandy.com/

Here is the classic video of the experiment done on the moon by Commander David Scott during the Apollo 15 moon mission in 1971. In the video, Scott drops a hammer and a feather on the virtually airless moon:

http://nssdc.gsfc.nasa.gov/planetary/lunar/apollo_15_feather_drop.html

Here's a link to lecture notes from Professor Michael Fowler of the University of Virginia that includes translated sections of Galileo's influential work on motion, *Dialogue Concerning the Two Chief World Systems,* also called the *Two New Sciences,* which was published in 1638:

http://galileoandeinstein.phys.virginia.edu/

For more background on Galileo's experiments on falling bodies and procedures for re-creating these experiments, visit this site:

http://www.juliantrubin.com/bigten/ galileofallingbodies.html

The always helpful Physics Classroom website provides lots of helpful equations and explanations:

http://www.physicsclassroom.com/class/1dkin/ u1l5b.cfm

http://www.physicsclassroom.com/class/circles/ u6l3e.cfm

Cornell University's Curious About Astronomy? uses a question and answer format to answer questions about physics and astronomy:

http://curious.astro.cornell.edu/question .php?number=465

http://curious.astro.cornell.edu/question .php?number=310

Chapter 4–Getting Heavy: Weight

HowStuffWorks is an excellent resource that I often turn to for information; this page describes how a bathroom scale works:

http://www.howstuffworks.com/inside-scale.htm

And here's a website that discusses how bathroom scales are going high tech:

http://www.npr.org/2011/03/21/134742980/ Bathroom-Scales-Theres-An-App-For-That

These three sites discuss mass and weight and other information about forces:

http://www.physicsclassroom.com/class/ newtlaws/u2l2b.cfm

http://www.edinformatics.com/math_science/ mass_weight.htm

The kilogram as a unit of mass has an interesting history; see

http://www.wired.com/magazine/2011/09/ ff_kilogram/all/1)

Gravity

I talked about gravity in this chapter. Here's a HowStuffWorks about gravity:

http://science.howstuffworks.com/environmental/ earth/geophysics/question232.htm

If you want to learn more about Newton's Law of Universal Gravitation, the idea that any two masses attract each other, based on the masses of the two objects and the square of the distance between both objects, you can find more at these websites:

http://csep10.phys.utk.edu/astr161/lect/history/ newtongrav.html

http://www.physicsclassroom.com/class/circles/ u6l3c.cfm

Albert Einstein (1879–1955) framed his views of gravity as the curvature of space and time. Objects are pulled into this curvature and this is called weight. You can find more discussion of Einstein's view of gravity on these websites:

http://www.pbs.org/wgbh/nova/physics/relativity- and-the-cosmos.html

http://www.einstein-online.info/elementary/ generalRT/GeomGravity

Here are a few fun simulations on gravity from the PHET site:

http://phet.colorado.edu/en/simulation/lunar- lander

http://phet.colorado.edu/en/simulation/gravity- force-lab

http://phet.colorado.edu/en/simulation/gravity- and-orbits

Weights and Scales

Many sources are available for masses, weight sets, spring scales, balances, and so on. I often use the following companies to order these products:

- **Cynmar** *http://www.cynmar.com/*
  - **Set of hook masses** *http://www.cynmar.com/ProductDetail/08011747_Hooked-Weight-Set-Contains-9-Weights-Wcase*
  - **Slotted weight set** *http://www.cynmar.com/ProductDetail/08011742_12-Pc-Slotted-Weight-Set-Whanger*
- **Arbor Scientific** *http://www.arborsci.com/*
  - **Spring scales (set for $35.00)** *http://www.arborsci.com/Products_Pages/Measurement/SpringScales.aspx*
- **Pasco** *http://www.pasco.com/*
  - **Spring scales, masses, balances** *http://www.pasco.com/products/basket/groups.cfm?&DID=9&groupID=347&start=1*
- **eNasco** *http://www.enasco.com/*
  - **Set of plastic stackable masses** *http://www.enasco.com/c/science/Measurement/Mass+Weight+Sets/Plastic+Stacking+Masses/*

I found digital electronic and computerized probes (motion sensors, force probes, force plates, and so on) from Vernier (*http://www.vernier.com/*). Pasco makes and sells probes, too.

You can find inexpensive food scales at stores like Wal-Mart; for example, search for **The Biggest Loser Digital Food Scale** at *http://www.walmart.com/*.

For larger masses, I often use barbell and dumbbell masses (weights). You may have some of these at home that you can use or have access to some in your school gym.

Trebuchets

Kelvin has several trebuchet kits (number 841861):

http://www.kelvin.com/Merchant2/merchant.mv?Screen=PROD&Product_Code=841861&Category_Code=ENDECD&Product_Count=6

I have built several trebuchets besides the ruler trebuchet discussed in this chapter. You'll find several videos on the website for this book.

You can find several other websites about building trebuchets. Here is a small selection:

http://blip.tv/make/make-your-own-trebuchets-make-video-podcast-146017

http://www.redstoneprojects.com/trebuchetstore/trebuchet_plans.html

http://www.instructables.com/ (search for **trebuchets**)

Here's a link to a PBS NOVA show on the construction of a huge trebuchet from the Middle Ages:

http://www.pbs.org/wgbh/nova/lostempires/trebuchet/

And finally the website for the *Punkin Chunkin* TV show on the Science channel:

http://science.discovery.com/tv/punkin-chunkin/

Chapter 5—Storming the Castle: Projectiles

You can find a translation of Galileo's book *Discourses and Mathematical Demonstrations Relating to Two New Sciences* online at this site:

http://ebooks.adelaide.edu.au/g/galileo/dialogues/index.html

Here are some additional websites involving projectiles and Galileo:

http://galileo.rice.edu/lib/student_work/experiment95/paraintr.html

http://www.mcm.edu/academic/galileo/ars/arshtml/mathofmotion2.html

Simulations

Here are four web-based projectile motion simulations (applets):

http://galileo.phys.virginia.edu/classes/109N/ more_stuff/Applets/ProjectileMotion/jarapplet .html

http://wps.aw.com/wps/media/ objects/877/898586/topics/topic01.pdf

http://phet.colorado.edu/en/simulation/ projectile-motion

http://www.walter-fendt.de/ph14e/projectile.htm

Math and Projectiles

Trigonometry is important in the study of physics and the analysis of many different types of problems including projectiles. Here are a few helpful websites:

http://www.clarku.edu/~djoyce/trig/

http://www.concordacademy.org/academics/ PrimerTrigPhysics.pdf

http://www.youtube.com/watch?v=LZQuGrDEJ4g

http://www4.ncsu.edu/unity/lockers/users/f/ felder/public/kenny/papers/trig.html

Catapult Kits

In the chapter, I built a catapult from scratch. You can also buy kits that have everything you need. Kelvin is a good source:

www.kelvin.com

You can find the rubber band catapult here:

http://www.kelvin.com/Merchant2/ merchant.mv?Screen=PROD&Product_ Code=841362&Category_ Code=ENDECD&Product_Count=0)

And the mousetrap catapult here:

http://www.kelvin.com/Merchant2/ merchant.mv?Screen=PROD&Product_ Code=841882&Category_ Code=ENDECD&Product_Count=3

You can find many websites about building catapults:

http://www.stormthecastle.com/catapult/how-to-build-a-catapult.htm

http://www.youtube.com/ watch?v=EVzorAMTnFM

http://www.knightforhire.com/catapult.htm

http://www.wikihow.com/Build-a-Desktop-Catapult

http://science.howstuffworks.com/transport/ engines-equipment/question127.htm

Chapter 6–Acceleration: Newton's Laws of Motion

Many television shows have been made about or refer to Isaac Newton. "Newton's Dark Secrets" is a NOVA show that originally aired on November 15, 2005, on PBS:

http://www.pbs.org/wgbh/nova/physics/newton-dark-secrets.html

http://www.pbs.org/wgbh/nova/newton/

The following websites are helpful introductions to Newton's Laws:

http://www.physicsclassroom.com/class/ newtlaws/

http://www.can-do.com/uci/ssi2001/ newtonsphysicsnb.html

You can find many biographies about Newton's life online:

http://www.clas.ufl.edu/users/ufhatch/pages/01-courses/current-courses/08sr-newton.htm

http://galileo.phys.virginia.edu/classes/109N/ lectures/newton.html

http://www.gap-system.org/~history/Biographies/ Newton.html

http://scienceworld.wolfram.com/biography/ Newton.html

The portrait of Isaac Newton on page 83 is by Godfrey Kneller and was painted in 1689.

Newton wrote several books during his lifetime. His most famous book, written in Latin, is known as *The Principia* (or *Mathematical Principles of Natural Philosophy*). In this book, he lays out his ideas of force and motion. The first website below offers a PDF version of the book; you can also find a current translation that you can purchase on Amazon.com:

http://www.new-science-theory.com/newton-principia.pdf

Here are a few websites about Galileo and the Law of Inertia (Newton's first law of motion):

http://zonalandeducation.com/mstm/physics/mechanics/forces/galileo/galileoInertia.html

http://csep10.phys.utk.edu/astr161/lect/history/galileo.html

More Experiments

You can perform lots of experiments using Newton's laws. These websites have additional suggestions:

http://sciencespot.net/Media/newtonlab.pdf

http://swift.sonoma.edu/education/newton/newton_1/html/newton1.html

http://www.pa.uky.edu/~ellis/Instr_Labs/Experiment_5_Newtons_Laws_1.pdf

http://thehappyscientist.com/science-experiment/newtons-laws

http://teachers.net/lessons/posts/661.html

To learn more about Newton's third law, you can perform several activities with balloons:

http://www.sciencebob.com/experiments/balloonrocket.php

http://exploration.grc.nasa.gov/education/rocket/BottleRocket/Shari/propulsion_act.htm

http://swift.sonoma.edu/education/newton/newton_3/html/newton3.html

http://www.youtube.com/watch?v=cec0y_8V1GE

http://www.education.com/science-fair/article/create-balloon-rocket/

You can also buy several toys—balloon helicopters and balloon cars—that rely on balloons and Newton's third law. Search for **"helicopters and balloons"** or **"cars and balloons"**.

Here is a good website with a more complete explanation of surface friction:

http://regentsprep.org/regents/physics/phys01/friction/default.htm

Water Rocket Kits and Launchers

I built my water rocket using information from the following website:

http://tclauset.org/21_BtlRockets/BTL.html

I purchased a launcher kit from Amazon.com.

You can find lots of websites about water rockets and water rocket launchers:

http://exploration.grc.nasa.gov/education/rocket/BottleRocket/about.htm

http://www.grc.nasa.gov/WWW/K-12/airplane/thrust1.html

http://www.nasa.gov/audience/foreducators/topnav/materials/listbytype/Water_Rocket_Construction.html

http://en.wikipedia.org/wiki/Water_rocket

http://www.instructables.com/id/Soda-Bottle-Water-Rocket/

http://www.sciencetoymaker.org/waterRocket/buildWaterRocketLauncher.htm

http://www.sciencetoymaker.org/waterRocket/index.htm

http://www.sciencetoymaker.org/waterRocket/greatLinks.htm

Here's a website about safety concerns and water rockets:

http://www.sciencetoymaker.org/waterRocket/safetyWaterRocket.htm

Chapter 7—Moving Forward: Kinetic Energy

Here are some websites about energy:

http://www.physicsclassroom.com/class/energy/

http://hyperphysics.phy-astr.gsu.edu/hbase/ enecon.html

Elastic Energy

As mentioned in the chapter, elastic energy can be found by taking the area of the force-distance graph. Not discussed in the text is the actual equation derived from this area. Elastic energy = $1/2 \times (k) \times (x^2)$, where k is the *spring constant* for the elastic substance in Newtons per meter, and x represents the distance pulled in meters. The spring constant is a measure of the elasticity of the substance.

The equation for elastic energy is applicable if the elastic substance has a stretch proportional to the distance pulled. These substances are called *Hookean* or *linear-elastic*. This idea was explored by Robert Hooke (1635–1703), a British scientist who researched elasticity. Often springs and rubber bands lose this proportionality if stretched too far. For example, a slinky, if stretched too hard, will permanently deform and not stretch back. In engineering terms, this is the yield point of the substance.

More information about elasticity can be found on the following websites:

http://hyperphysics.phy-astr.gsu.edu/hbase/ permot2.html

http://www.animations.physics.unsw.edu.au/jw/ elasticity.htm

http://www.madphysics.com/exp/hysteresis_and_ rubber_bands.htm

http://prettygoodphysics.wikispaces .com/file/view/Chapter+10+- +Elasticity+%26+Oscillations.pdf

Mousetrap-Powered Car

You can find many online sources for mousetrap-powered cars. Here are a few:

http://www.scienceguy.org/Articles/ MousetrapCars.aspx

http://www.docfizzix.com/

http://www.pbs.org/saf/1208/teaching/teaching .htm

http://gspyda.hubpages.com/hub/How-to-Build- a-Mouse-Trap-Car---A-Step-by-Step-Guide

Chapter 8—Whacks and Bangs: The Physics of Collisions

Here are a few extra resources and websites about momentum and collisions:

http://theory.uwinnipeg.ca/physics/mom/index .html

http://www.physicsclassroom.com/class/ momentum/u4l1b.cfm

http://www.youtube.com/watch?v=vQelPiezm9k

http://www.youtube.com/watch?v=cvfB6y_n_QA

http://www.animations.physics.unsw.edu.au/jw/ momentum.html

The following are computer simulations of collisions that allow you to change the mass and speeds of different balls or colliding objects:

http://phet.colorado.edu/en/simulation/collision- lab

http://www.walter-fendt.de/ph14e/collision.htm

http://www.myphysicslab.com/collision.html

Rocket Engines

I obtained my rocket engines from a local hobby store. They are generally in packages of three or four and come with igniters and plugs. The engines I used were manufactured by Estes. Estes and other model rocket companies have extensive websites. Here are several websites that give engine information and specifications:

http://www.hobbylinc.com/rockets/info/rockets_ enginefacts.htm

http://www.estesrockets.com/

http://www.rocketarium.com/

http://www.apogeerockets.com/

http://www.aerotech-rocketry.com/

http://www.lunar.org/docs/handbook/motors .shtml

 BE CAREFUL!

You can also find websites about buying different chemicals and building your own rocket engines. Make sure to research this area and understand the safety concerns of mixing your own rocket fuel.

In terms of the engine launching ignition system, I used an Estes Electron Beam. The Electron Beam model can be purchased for around $15. Building your own is a bit more expensive but a nice project that you'll work on in Chapter 9.

Using a Vernier (http://*www.vernier.com/*) or Pasco (http://*www.pasco.com/*) force probe to test the impulse of rocket engines typically gives good results. This technique has been written about in other lab manuals and on other websites; however, the needed hardware and software is expensive. This method requires an engine bracket, force probe, interface box, computer, and software. Pasco has an engine bracket that you can purchase. Here are sites with additional information:

http://www.nsa.gov/academia/_files/collected_ learning/high_school/science/model_rockets.pdf

http://www.physicssource.ca/pgs/3005_mom_ elab_23.html

http://arxiv.org/ftp/arxiv/papers/0903/0903.1555 .pdf

Here is the citation for the article in *The Physics Teacher*, "Measuring Model Rocket Engine Thrust Curves" by authors Kim Penn and William V. Slaton, which is discussed in the chapter:

http://www.eric.ed.gov/ERICWebPortal/search/ detailmini.jsp?_nfpb=true&_&ERICExtSearch_ SearchValue_0=EJ912903&ERICExtSearch_ SearchType_0=no&accno=EJ912903

NOTE

The scientist for this chapter is Dr. John Stapp. I found researching his life fascinating. If you find yourself in the Alamorgordo, New Mexico, area, there is the Museum of Space History with a special exhibit honoring Dr. Stapp: *http://www.nmspacemuseum.org/*.

Chapter 9–Blast Off! The Physics of Rocketry

You can find an incredible amount of information concerning model rocketry on the Internet. Many model rocket clubs exist, too, and there may be one close to you. The science and mathematics behind rockets and their motion are complicated. By ignoring air resistance and neglecting the change of mass, you can take a complicated motion and simplify the equations to come up with estimates of the rocket's altitude, speeds, and acceleration. If you would like to study a complicated analysis, however, here are a few websites with additional information to get you started:

http://en.wikipedia.org/wiki/Tsiolkovsky_rocket_ equation

http://exploration.grc.nasa.gov/education/rocket/ rktpow.html

http://www.rocketmime.com/rockets/rckt_eqn .html

Additional resources concerning rocket launching ignition systems, launch pads, and rockets can be found here:

http://home.earthlink.net/~rbogerjr/

http://www.robotroom.com/Model-Rocket-Launch-Controller.html

http://www.instructables.com/id/remote-model-rocket-ignitor/

http://www.youtube.com/watch?v=Dqx_GpqcTck&feature=related

http://www.youtube.com/watch?v=yurohmsmY_Y&feature=relmfu (see also parts 2 and 3)

Several books about model rocketry are available:

Handbook of Model Rocketry, 7th Edition, G. Harry Stine and Bill Stine

Modern High-Power Rocketry 2, by Mike Canepa

Inclinometers

A physical method of estimating rocket height is with an inclinometer. They can be purchased (Estes Altitrak, *http://www.estesrockets.com/302232-altitraktm*) or constructed. There are many uses for inclinometers, including in various professions.

There are some inclinometer apps, too, for mobile phones. Here are a few websites with additional information:

http://www.exploratorium.edu/math_explorer/howHigh_makeInclino.html

http://www.youtube.com/watch?v=GMLcU1Qknts

http://itunes.apple.com/us/app/advanced-level-inclinometer/id288338285?mt=8

Here is more information about the trigonometry used (for the inclinometer):

http://www.sigmarockets.com/blog/2011/08/mathematics-in-motion-how-high-did-my-rocket-go/

http://www.sciencebuddies.org/science-fair-projects/project_ideas/Math_p026.shtml

Here is some information on flights for the Baby Bertha Estes rocket:

http://www.rocketreviews.com/estes-industries-baby-bertha.html

http://www.estesrockets.com/001261-baby-berthatm

http://www.youtube.com/watch?v=ZniLqBb1Qog

GLOSSARY

acceleration A physics term meaning the rate at which an object speeds up or slows down with respect to time. Positive acceleration results in an object getting faster; negative acceleration is when an object slows down.

acceleration due to gravity The numerical value of the constant rate of acceleration from gravity, symbolized by a lowercase g. On Earth, g is (on average) 9.8 meters per second per second, 980 centimeters per second per second, or 32 feet per second per second.

air resistance The force on an object due to the influence of air molecules on the object as it moves. *See also* drag.

apex The highest point of an object's trajectory or path. For an object in free fall upward and then downward, the apex is the highest point achieved by the object and the position where the object has no speed.

average A single number that represents a group of numbers (data). Add the data and divide by the number of trials to obtain the average. You often take numerical averages in science.

ballistics The study of the motion of projectiles often through firearms, siege engines, and cannons.

bouncy A type of collision, also called an *elastic collision*. In a perfectly elastic collision both kinetic energy and momentum are conserved. *See also* elastic energy.

calibrate Instruments that measure weight and forces, such as bathroom scales, often need to be calibrated. Calibration involves measuring known weights and then placing these known values on the instrument. After the instrument is calibrated, unknown weights can be measured. If an instrument is not calibrated correctly, measurements will be incorrect.

collisions One or more objects hitting another object(s).

conclusions The answers you get after experiments are completed.

conservation of momentum Momentum of objects before a collision is equal to the momentum of objects after a collision. The conservation of momentum is true regardless of the type of collision if you ignore any frictional effects or other external forces.

constant or uniform acceleration An acceleration that stays constant through a time interval.

constant speed Going the same speed or distance per unit of time.

contact forces   Forces that express themselves through the contact of two objects. Friction, tension, and the normal force are types of contact forces.

data   Information, often numbers, that comes from conducting experiments.

density   Defined as the mass of an object divided by its volume. Something that is more dense can be thought of as more compact or concentrated in nature.

distance   How far an object moves.

distance-time graph   A graph that shows you how far something moves in a given amount of time. Sometimes called a *position-time graph*.

drag   The force of air resistance. The impact of the atmosphere on a moving object depends on the object's speed. *See also* air resistance.

elastic   A type of collision where momentum and kinetic energy are conserved.

elastic energy   Energy trapped in an elastic substance such as a rubber band, spring, or bungee cord. You can find this energy by measuring the force needed to pull the substance a certain distance and then finding the area of the graph formed. Elastic energy is also called *spring energy* and is given the symbol, E_{el}, and is measured in *Joules* (J).

elasticity   The stretching of a substance.

energy conservation (Law of Energy Conservation)   The concept that energy cannot be created or destroyed. It can, however, change forms.

energy transformation   The changing of forms for a type of energy. For example, an object thrown directly upward will have its kinetic energy change into gravitational potential energy as it slows down and gains more height.

equation   An equation relates numbers and concepts and usually sets up an "equals" statement. Scientists often try to come up with an equation from an experiment. For example, when an object stays at one speed, you can say that this object is moving at a constant speed. This speed is the distance moved divided by the time it moved. You can use the equation *distance/time = constant speed*.

experiment   How physicists learn about our world and universe. An experiment is a repeatable procedure designed to test a hypothesis or idea. *See also* hypothesis.

field forces   When the object that generates the force does not have to come in physical contact with other objects to exert that force. Field forces are expressed as action-at-a-distance forces. Gravitational force is a field force.

force   An interaction between two objects. A force is commonly defined as a push or a pull. Force is measured in Newtons (metric system) or pounds (US).

force diagram   A diagram of an object showing the forces on an object. See also *free body diagram*.

force-time graph   A graph with time on the x axis and force on the y axis. The area under this graph is called the *impulse*.

free body diagram   A diagram of an object showing the forces on an object. Also called a *force diagram*.

free fall   Free fall is the motion of an object as it falls under the effects of a planet's gravity.

GRACE GRACE or the *Gravity Recovery and Climate Experiment* is an experiment NASA launched in 2002, involving a pair of satellites. One of the goals of the mission was to measure the small variations of our planet's gravity.

gravitational potential energy (GPE) Energy that an object has based on its position. The equation to find GPE is a product of the object's mass (m in kg), the acceleration due to gravity ($g = 9.8$ m/s/s for the Earth), and the object's height above a reference line or baseline (h in meters), GPE = $m \times g \times h$. Gravitational potential energy is also measured in Joules (J).

gravity One of the fundamental forces of the universe, gravity is the attraction between objects that have mass. Gravity causes free fall motion. The force that pulls on objects and gives objects their weight.

Hooke's Law A relationship discovered by Robert Hooke (1635–1703). A substance that follows Hooke's Law stretches proportionally as weight is applied. For example, if 10 Newtons of force is applied to a spring, it might stretch 5 centimeters. If 20 Newtons of force is applied to the spring, it would then stretch 10 centimeters.

hypothesis Statement or question that can be tested by conducting experiments.

impulse The product of force and time.

impulse-momentum theorem The theorem or equation that says impulse is equal to the change of momentum.

inelastic A type of collision where momentum but not kinetic energy is conserved.

inertia An object's tendency to maintain its current state of motion and resist changes in its state of motion. A heavier object has more inertia that a lighter object and has a greater tendency to maintain its state of motion more than a lighter object.

kilogram A metric unit of mass. *See also* mass.

kinematics The study of the equations and graphs of moving objects without considering the forces acting on the object to create the motion.

kinetic energy (KE) Energy that a moving object contains. Kinetic energy is based on the mass, m, of the object in kilograms, and the velocity or speed of the object measured in meters/second. The equation to find kinetic energy is $\frac{1}{2} * (mass) \times (speed^2)$. Energy is measured in units called Joules (J).

law A scientific law generalizes a set of observations and is often mathematical in nature.

lever arm In a catapult or trebuchet, the part of the structure involved with projecting the object. Usually this arm rotates in some way due to the force of gravity or with elastic or tension forces causing the object to be moved forward and upward.

macroscopic observations Observations that are large enough to be seen by the naked eye.

mass The amount of matter in a physical object.

matter The physical stuff around you. The universe can be thought of as being made up of matter and energy. Matter is made up of different states or arrangements; gas, liquid, and solid are three states of matter.

microscopic observations Observations so small that you need microscopes or other tools to see them.

momentum Mass times speed (velocity), usually given in units of kg*m/sec.

net force   An overall sum of forces. Can be zero or some non-zero number. If the net force equals 0, the forces are balanced and the object will not accelerate.

Newton   A unit of force in the metric system.

Newton's first law   The speed (velocity) of an object will stay constant unless a nonzero net force (unbalanced force) acts on the object. This law has several parts. If the object is not moving, it remains motionless unless an unbalanced force acts on it. If the object is moving at a constant speed, then an unbalanced force is needed to change this velocity. Newton's first law is often stated as follows: a body in motion tends to stay in motion; a body at rest tends to stay at rest.

Newton's second law   A nonzero net force (unbalanced force) acting on an object accelerates the object, or $F_{net} = mass \times acceleration$.

Newton's third law   For every force, there is an equal and opposite force. For every action, there is an equal and opposite reaction. Forces are interactions between two objects.

normal   The force supporting weight from a floor, desk, table, or other surface.

oxidizer   A substance that combines with the fuel in a jet or rocket engine.

parabola (parabolic relationship)   A parabola is a special curve seen on a distance-time graph that shows an acceleration. A parabola on a distance-time graph suggests an object does not move the same amount of distance during successive time intervals. The distance moved is increasing and the object is getting faster (positive acceleration, parabola directed upward or upward opening) or decreasing and getting slower (negative acceleration, parabola directed downward or downward opening).

perpendicular   An object that is at a right angle (90 degrees) to another object is said to be perpendicular.

pounds   A unit of force used in the United States. There are 4.45 Newtons per pound.

prediction   In science, equations are used to predict or to answer where an object might be in the future.

projectile   An object projected or thrown that moves in a trajectory under the influence of gravity. Typically, a projectile has vertical and horizontal motions that occur simultaneously to form a trajectory.

projectile motion   Motion that has horizontal and vertical parts that occur together to form a parabolic trajectory. Ignoring air resistance, horizontally the projectile moves at a constant speed; vertically, the object is in free fall.

recoil   An interaction where one object splits apart. Momentum is conserved in these interactions too.

reference line (baseline)   To obtain an object's gravitational potential energy, assign a reference line for height measurements. Usually this is taken from the ground or floor. Height is 0 at the reference line. For some situations, it can help to assign a different reference line than the ground.

siege engines A weapon of war generally designed to launch projectiles at castles.

slope The steepness of a line on a graph. The slope on a distance-time graph (position-time graph) represents the speed (velocity) of the object. The slope on a speed-time graph (velocity-time graph) represents the acceleration of an object.

speed Distance divided by time, or "how fast" an object moves. Sometimes called *velocity,* although the terms differ technically. *Speed* is distance divided by time, whereas *velocity* is displacement divided by time. *Displacement* is calculated by taking a final position minus an initial position. (For this book, I use the terms *speed* and *velocity* synonymously.)

speed-time graph A graph that plots an object's speed on the y axis and time on the x axis. Horizontal lines on the speed-time graph occur when objects maintain one speed or velocity. With constant acceleration, you get a diagonal line on the speed-time graph. Also called a *velocity-time graph.*

sticky Example of an inelastic collision whereby the two objects stick together after a collision and move at some speed as one unit.

surface friction A resistive force arising from the motion of one surface relative to, or across, another surface.

tension A tightly pulled string, rope, or cable. Often supports weight.

theory A theory summarizes one or more hypotheses and has been rigorously tested many times.

thrust A force found in balloons, rockets, and jet engines that pushes the balloon, rocket, or jet forward.

trajectory The path followed by a projectile.

trendline A pattern that can be formed on a graph that "follows" the data. This pattern may be linear (a diagonal line) or a curve of some type (a parabola, for example). Some math textbooks call this the *line of best fit.* Using mathematical techniques or computer processes, you can generate an equation from a trendline.

trial Repeating an experiment.

two-dimensional motion Motion along two axes simultaneously. Projectile motion is an example of two-dimensional motion.

unbalanced/balanced forces Forces acting on an object can be balanced or unbalanced. Balanced forces mean that all forces sum to equal zero and the object does not accelerate. Unbalanced forces do not sum to equal zero, and consequently, the object will accelerate.

vacuum The state of no atmosphere or when the effects of air resistance are negated.

vector In science, a vector is a concept that needs an amount and a direction. Velocity, acceleration, and force are all vector quantities.

weight A force due to the effects of gravity ($W = mg$).

Index